T0130942

SDH/SONET Explained in Functional Models

SDH/SONET Explained in Functional Models

Modeling the Optical Transport Network

Huub van Helvoort
Networking Consultant, the Netherlands

John Wiley & Sons, Ltd

Telephone (+44) 1243 779777

Email (for orders and customer service enquiries): cs-books@wiley.co.uk
Visit our Home Page on www.wiley.com

Other Wiley Editorial Offices

John Wiley & Sons Inc., 111 River Street, Hoboken, NJ 07030, USA

Jossey-Bass, 989 Market Street, San Francisco, CA 94103-1741, USA

Wiley-VCH Verlag GmbH, Boschstr. 12, D-69469 Weinheim, Germany

John Wiley & Sons Australia Ltd, 42 McDougall Street, Milton, Queensland 4064, Australia

John Wiley & Sons (Asia) Pte Ltd, 2 Clementi Loop #02-01, Jin Xing Distripark, Singapore 129809

John Wiley & Sons Canada Ltd, 22 Worcester Road, Etobicoke, Ontario, Canada M9W 1L1

Wiley also publishes its books in a variety of electronic formats. Some content that appears in print may not be available in electronic books.

British Library Cataloguing in Publication Data

A catalogue record for this book is available from the British Library

ISBN 0-470-09123-1

Typeset in 10/12pt Palatino by Thomson Press (India) Limited, New Delhi, India.
Printed and bound in Great Britain by TJ International, Padstow, Cornwall.
This book is printed on acid-free paper responsibly manufactured from sustainable forestry in which at least two trees are planted for each one used for paper production.

To my wife Leontine for her support and patience,
and in memory of my parents

Contents

Preface

The use of a natural language to describe the functionality in transmission networks and transport equipment will lead to misinterpretation of the written requirements and cause equipment not to interoperate. The growth in complexity of the functionality and diversity of the optical transport network capabilities to be described, and the number of different users, for example, system engineers, marketing, customers, developers, standards representatives, meant that it was necessary to develop and define a common language.

In this book I describe this language, i.e. the methodology that is used to model the functionality of transport networks and transport equipment. The functional modeling methodology is applicable in connection-oriented networks, e.g. PDH, SDH, SONET, OTN, as well as connectionless networks, e.g. Ethernet, MPLS. The emphasis in this book is on the explanation of the functional modeling methodology and its use as a description tool. Examples are provided to help the reader in understanding modeling technique.

Based on my experience with the use of functional models over the past ten years, I expect that many readers of this book will be System Engineers and Functional Architects who are employed by Optical Transport Network operators, Optical Transport equipment manufacturers and device manufacturers, especially those who are responsible for transport-related functionality at Networking or Network Element level. It will help them to use and develop functional models in the area of their responsibility.

I assume that optical network, equipment and device development engineers as well as system verification, system test and interoperability test engineers will use this book as a guideline.

Finally, I hope that this book will be used by students in telecommunications technology and by members of the IEEE community as a reference to acquire the skill of functional modeling.

Acknowledgements

I especially thank Eve Varma, Maarten Vissers and George Newsome for their cooperation and for sharing my enthusiasm.

Huub van Helvoort, M.S.E.E., Senior member IEEE.

Abbreviations

AcSQ	Accepted Sequence number
ADM	Add-Drop Multiplexer
AI	Adapted Information
AIS	Alarm Indication Signal (i.e. Alarm Inhibit Signal)
ANSI	American National Standards Institute
AP	Access Point
API	Access Point Identifier
APS	Automatic Protection Switch
ATM	Asynchronous Transport Module
AU	Administrative Unit
BP	Bridge Protocol
CBRx	Constant BitRate signal with approximate bitrate x
CCAT	Contiguous conCATenation
CCITT	Comite Consulatif Telegraphique et Telephonique (now ITU-T)
CI	Characteristic Information
CP	Connection Point
CRC	Cyclic Redundancy Check
DCC	Data Communications Channel
DCN	Data Communications Network
DXC	Digital Cross-Connect
EOW	Engineer Order Wire
ETSI	European Telecommunications Standards Institute
FCS	Frame Check Sequence
FD	Flow Domain
FDFr	Flow Domain Fragment
FEBE	Far End Block Error
FERF	Far End Receive Failure
FOP	Failure of Protocol

FP	Flow Point
FPP	Flow Point Pool
GFP	Generic Framing Process
IEEE	Institute of Electrical and Electronics Engineers
IP	Internet Protocol
ISO	International Organization for Standardization
ITU-T	International Telecommunications Union—Telecommunication Standardization Sector (former CCITT)
LAN	Local Area Network
LC	Link Connection
LCAS	Link Capacity Adjustment Scheme
LOA	Loss of Alignment
LOM	Loss of Multi-frame
MAC	Media Access Control
MAN	Metro Area Network
MI	Management Information
MND	Member Not De-skewable
MP	Management Point
MPLS	Multi-Protocol Label Switching
MS-SPRing	Multiplex Section—Shared Protection Ring
MSn	Multiplex Section of level n
MSAP	Multi-Service Access Platform
MSPP	Multi-Service Provisioning Platform
MSSP	Multi-Service Switching Platform
MSTP	Multi-Service Transport Platform
MSU	Member Service Unavailable
MTBF	Mean Time Between Failures
MTTR	Mean Time To Repair
NC	Network Connection
NNI	Network to Network Interface
NUT	Non-pre-emptible Unprotected Traffic
OAM/OA&M	Operation Administration and Maintenance
OSI	Open Systems Interconnection
OOS	OTM Overhead Signal
OSn	Optical Section of level n
OTM	Optical Transport Module
OTN	Optical Transport Network
PCI	Protocol Control Information
PDH	Plesiochronous Digital Hierarchy
PLC	Partial Loss of payload Capacity
POH	Path OverHead

PRC	Primary Reference Clock
QoS	Quality of Service
RDI	Remote Defect Indication
REI	Remote Error Indication
RI	Remote Information
RP	Remote Point
RSn	Regenerator Section of level n
Sk	Sink
So	Source
SAN	Storage Area Network
SD	Signal Degrade
SDH	Synchronous Digital Hierarchy
SDU	Service Data Unit
SF	Signal Fail
SLA	Service Level Agreement
SNC	Sub-Network Connection
SNCP	Sub-Network Connection Protection
SNC/I	SNCP using Inherent monitoring
SNC/N	SNCP using Non-intrusive monitoring
SNC/S	SNCP using Sub-layering
SONET	Synchronous Optical Network
SQ	Sequence number
SQM	SQ Mismatch
SSD	Server Signal Degraded
SSF	Server Signal Failed
SSM	Synchronization Status Messaging
STM-N	Synchronous Transport Module (level) N
TCM	Tandem Connection Monitoring
TCP	Termination Connection Point
TFP	Termination Flow Point
TI	Timing Information
TLC	Total Loss of payload Capacity
TP	Timing Point
TSD	Trail Signal Degraded
TSF	Trail Signal Fail
TTI	Trail Trace Identifier
TU	Tributary Unit
UNI	User to Network Interface
VC–n	Virtual container (level) n
VCAT	Virtual conCATenation
VCG	Virtual Concatenation Group
WAN	Wide Area Network

1

Introduction

1.1 HISTORY

1

Introduction

A telecommunications network is a complex network that can be described in a number of different ways depending on the particular purpose of the description. In this book the optical transport network will be described as a network from the viewpoint of the capability to transfer information. More specifically, the functional and structural architecture of optical transport networks is described independently of the networking technology, for example, distribution, platforms, packaging. The methodology used for this description is commonly referred to as functional modeling and is used in many standards documents to describe the functional architecture of existing and evolving PDH, SDH, OTN, ATM, Ethernet and MPLS networks. The functional model is also used extensively by operators to describe their network and by manufacturers to describe their equipment or devices.

1.1 HISTORY

The development of functional models for use in telecommunication networks was a combined effort of network operators and equipment manufacturers. After an extensive analysis of existing transport network structures, the functional modeling methodology was first introduced in the standards documents of the *European Telecommunications Standards Institute* (ETSI) around 1995. In the ETSI standards the methodology was used to model the SDH network and its equipment. After the introduction and standardization in ETSI, the *International Telecommunications Union – Telecommunication Standardization Sector*

SDH/SONET Explained in Functional Models Huub van Helvoort

(ITU-T) also adapted the functional modeling in 1997. Although initially used for the specification of SDH, later it was applied in the specification of other technologies. Currently, work is in progress on modeling Ethernet networks. There is an increased interest in the *American National Standards Institute* (ANSI) to adopt the functional model methodology in their standards.

1.2 JUSTIFICATION

There were several reasons to start the study and development of a methodology to model a transport network or equipment in a functional way. Some of these reasons were:

- *Increased complexity*. Owing to the natural growth of optical transport networks and equipment, the contained functionality increased as well.
- *Increased variety*. Owing to the growth in complexity, the number of required functions also increased as well as the number of possible combinations of these functions.
- *Multiple applications*. The same transport equipment is labeled differently, e.g. multiplexer, cross-connect, line system, depending on the application in the network topology.
- *Written requirements*. Generally, in a natural language, there have the following disadvantages:
  - no common language; requires translation
  - voluminous; easy to lose overview
  - inconsistent; often dependent on the writer's background
  - ambiguous; uses a natural language
  - incomplete.

These reasons meant that it became more and more difficult to manage the transport network and equipment. It was almost impossible to guarantee the compatibility and interoperability of equipment based solely on written documentation. Consequently, this created the need for a new language that showed similarities where networks and equipment were similar and differences where they were dissimilar.

Considering the reasons mentioned above, the following requirements were taken as input for the study to establish a new description methodology, i.e. a common language:

- It should provide a flexible description of the functional architecture at transport network level that takes into account varying partitioning and layering requirements.
- It should identify functional similarities and differences in heterogeneous technology-based layered transport network architecture.
- It should be able to produce network element functional models that are traceable to and reflective of network level requirements.
- It should establish a rigorous and consistent relationship between transport network functional architecture and management information models.

In addition, the established methodology should have the following characteristics:

- it is simple;
- it is short;
- it is visual;
- it contains basic elements;
- it provides combination rules;
- it supports generic usage;
- it has recursive structures;
- it is implementation independent;
- it is transport level independent;
- it has the capability to automate generation and verification.

The result of the study is the definition and standardization of the functional modeling methodology. With a functional model it is possible to present:

- Optical Transport Network capabilities, independent of actual deployed equipments.
- Transport Equipment capabilities, independent of actual equipment implementation.

The unambiguous specification produced by applying the methodology will provide a unique definition of transport networks and equipment towards:

- Optical Transport Network operators;
- Optical Transport Equipment and Device manufacturers;
- Network Management Systems and Element Management Systems.

1.3 REMARKS ON THE CONCEPT

The analysis and decomposition of existing transmission networks resulted in the definition of the atomic functions that are used in the functional modeling methodology. These atomic functions can be used to compose the functional models of the same existing, legacy and future transmission networks.

This concept is not new. Some other technologies that have used atomic models are listed below.

- Hardware (analog). The atomic functions used in this technology are, for exampl, resistors, capacitors, inductors, diodes, and transistors. These atomic functions can be used to model an analog circuit or network, for example, an amplifier, a cable or even a digital circuit like an OR gate as illustrated in Figure 1.1.
- Hardware (digital). The atomic functions utilized in this technology are, for example, AND, NAND, OR, NOR, INVERT and XOR gates. Even though these functions can be represented by the atomic functions of the analog hardware as shown in the previous example, in this technology they would provide too much detail and would make the description of the digital circuit too complex. Thus the gates are in fact compound functions representing the analog

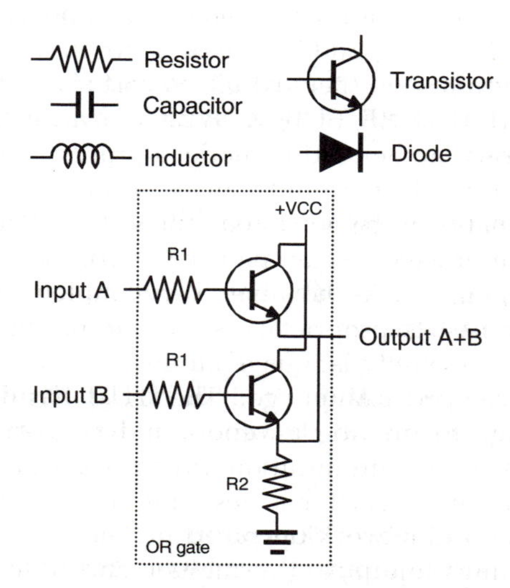

Figure 1.1 electric circuit symbols and example.

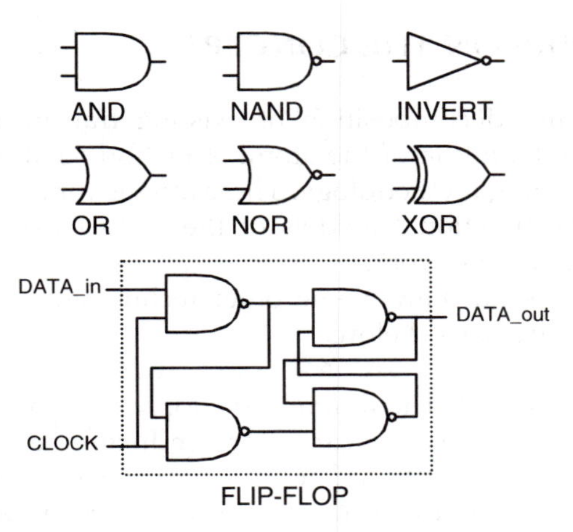

Figure 1.2 logic symbols and example.

hardware atomics. These atomic functions in this technology can again be used to model a digital circuit, for example, a flip-flop, which can be used as a compound function in other models, a microprocessor or a digital transmission function (e.g. a multiplexer, framer). Figure 1.2 shows the atomics and an example circuit.

- Software (assembly language). Even the primitives in textual form can be considered as building blocks to describe a particular function. Examples of the textual atomic functions are JUMP, JMPNC, LOAD, STORE, SUB, XOR and NOP instructions. These atomic functions can be used to model a process. Figure 1.3 depicts a simple example. Instructions can be grouped together to form procedures that can be CALLed and RETURNed from when finished; these procedures can be used as compound functions.

  Assembly language is however very implementation specific; every vendor has its proprietary set of atomic models and there are no generic assembly language atomic functions.
- Software (higher order language). The ITU-T has defined a higher order language to provide a vendor independent programming capability for telecommunication processes: CHILL, the CCITT Higher Level Language (for a description, see ITU-T Rec. Z.200, 1999). Another and more widespread higher order language is the C programming language (Kernighan and Ritchie, 1978). The atomic functions are, for example, FOR, WHILE, IF-THEN-ELSE,

```
;-----------------------------------------------------
; Read the input port, IPORT, 5 times, and store into
; 5 consecutive memory locations, elements of array
; DATA[i]. Also read each element of DATA[i] and
; write it to output port, OPORT.
;------------------------------------------------------
.EQU    IPORT      0               ; #define IPORT

.EQU    OPORT      1               ; #define OPORT

;

        XOR        R0, R0, R0      ; clear register R0

LBL0:   SUBL       NULL, R0, 5     ; as long as R0 < 5 {

        JMPNC      LBL1            ;

        LOADP      R1, IPORT       ; read input port into R1

        DSTOREM    DATA(R0), R1    ; store R1 in memory

        DLOADM     R2, DATA(R0)    ; read memory into R2

        STOREP     OPORT, R2       ; write R2 to output port

        ADDL       R0, R0, 1       ; increment R0

        JMP        LBL0            ; }


LBL1:   NOP                        ; end of program.
```

Figure 1.3 Assembly language example.

CASE, and ASSIGNMENT. Frequently used routines can be collected in a library and used as compound functions. These atomic functions can be used to model a process, for example, the same as the assembly language example above (see Figures 1.3 and 1.4.)

Higher order languages are independent of the implementation. There are, however, only a few (micro-) processors that can interpret this higher order language; a translator, or compiler, is used to generate the implementation specific assembly language understood by a particular (micro-) processor.

- Process descriptions using state diagrams. The atomic functions in this methodology (e.g. SDL Specification and Description Language, ITU-T Rec. Z.100, 2002) are STATE, INPUT, OUTPUT, TASK and DECISION. The TASK symbol may represent a procedure that

```
;-----------------------------------------------------------
; Read the input port 5 times, and store into
; 5 consecutive memory locations, elements of array
; DATA[i]. Also read each element of DATA[i] and
; write it to output.
;-----------------------------------------------------------
main()
{
        int c;
        int DATA[5];

        c = 0;

        while ( c < 5) {
        c = getchar();
        DATA[I] = c;
        putchar() = c;
        }
}
        /*end of program. */
```

Figure 1.4 C Language example.

can again be specified in SDL and can be considered as a compound function. These atomic functions can be used to describe a process, for example, subscriber signaling, Link Capacity Adjustment Scheme (LCAS, see ITU-T Rec. G.7042, 2004). Hardware and software designers can use these models when implementing a specific process. Figure 1.5 shows an example of the graphical representation: SDL/GR. This is a part of the SDL diagram describing the Source side processing in LCAS (see ITU-T Rec. G.7042, 2004).

SDL also has a textual phrase representation as shown in Figure 1.6: SDL/PR. Tools exist that use this text to generate the graphical representation and and/or generate executable code for testing purposes.

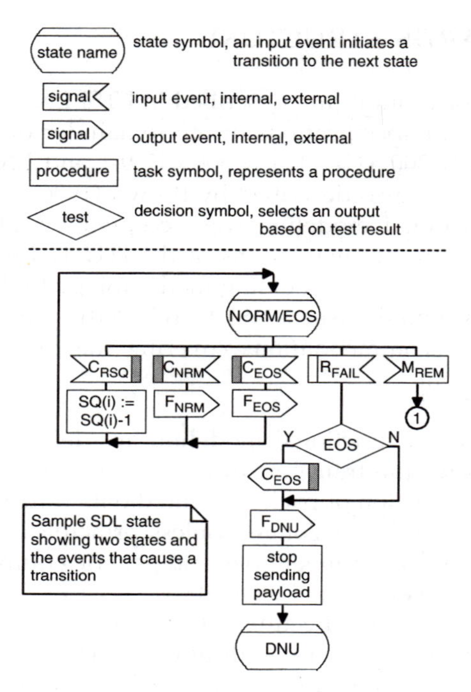

Figure 1.5 SDL/GR functional model example.

```
PROCESS LCAS_source_side;

        STATE NORM/EOS;
                INPUT CRSQ;
                        TASK 'SQ(i) := SQ(i) - 1';
                        NEXTSTATE NORM/EOS;
                INPUT CNRM;
                        OUTPUT FNRM;
                        NEXTSTATE NORM/EOS;
                INPUT CEOS;
                        OUTPUT FEOS;
                        NEXTSTATE NORM/EOS;
                INPUT RFAIL;
                        DECISION 'MBRST';
                        (EOS): OUTPUT CEOS;
                        (NORM): ;
                        ENDDECISION;
                        OUTPUT FDNU;
                        TASK'stop_sending_payload';
                        NEXTSTATE DNU;
                INPUT MREM;
                        .... ;
        STATE DNU;
                .... ;
ENDPROCESS;
```

Figure 1.6 SDL/TR example.

1.4 STANDARDS STRUCTURE

The modeling conventions are described in ETSI EN 300 417-1-1(2001) and the equipment specifications in the remainder of this series (EN 300 417-2-1 to EN 300 417-7-1, EN 300 417-9-1 and EN 300 417-10-1). The methodology is also described by Brown (1996).

After the functional modeling was accepted by ETSI it was also introduced in the recommendations of the ITU-T. The ANSI has not yet adopted the functional modeling methodology to describe SONET networks and equipment (see ANSI T1.105, 2001). Currently there is a whole suite of Recommendations covering the full functionality of network equipment:

- The principles of functional modeling are defined in ITU-T Rec. G.805 (2000) for the transport of connection oriented signals and, since the introduction of packet oriented data transport, ITU-T Rec. G.809 (2003) defines the connectionless principles.
- Functional modeling conventions and generic equipment functions are defined in ITU-T Rec. G.806 (2004).
- The SDH network architecture can be found in ITU-T Rec. G.803 (2000) and the equipment specification in ITU-T Rec. G.783 (2004).
- The OTN network architecture can be found in ITU-T Rec. G.872 (2001) and the equipment specification in ITU-T Rec. G.798 (2004).
- For PDH only the equipment specification is available in ITU-T Rec. G.705 (2000).
- The ATM network architecture is defined in ITU-T Rec. I.326 (1995) and the functional characteristics are described in ITU-T Rec. I.732 (2000).
- The Ethernet network architecture can be found in ITU-T Rec. G. 8010 (2004) and the equipment specification in ITU-T Rec. G. 8021 (2004).
- The MPLS network architecture can be found in ITU-T Rec. G. 8110 (2005) and the equipment specification in draft ITU-T Rec. G. mplseq (2005).
- Network and Network Element management functionality is described in ITU-T Rec. G.7710 (2001) for common equipment, in G.784 (1999) for SDH networks and in G.874 (2001) for OTN equipment. MPLS OAM functionality is defined in ITU-T Rec. Y 1710 (2002).

2

Functional modeling

Structuring the network

In a telecommunications network, various functions can be determined to describe the operation of the network. These functions can be classified into two distinct groups. One group is the transport functional group, i.e. functions required to transfer any telecommunications information from one or more points to one or more other points. The other group is the control functional group, i.e. functions required to provision, maintain and supervise the functions in the transport functional group. This book describes mainly the transport functional group.

The atomic functions that are used in the functional modeling methodology are derived from a thorough analysis and by decomposition of existing transmission networks. These functions can then be used to compose the functional models of the same existing and future transmission networks. While traditional networks are connection oriented, because they were designed to transport voice, the next generation networks will grow towards data transport and become connectionless.

2.1 FUNCTIONAL ARCHITECTURE OF TRANSPORT NETWORKS

The transfer of user information in a transport network from one location to another can be either bi-directional or uni-directional.

Functions in the transport functional group can also be used to transfer network control information, for example, signaling, operations and maintenance information, for the control functional group.

The existing transport network is a vast and complex network that contains many different components. In a telecommunication network there are, for example, switching systems, transmission systems, signaling systems and management systems. These systems, or network elements, are located in many nodes, connected by a complex mesh of links and heavily interactive. A system contains many functions, for example, a transmission system will have framing, multiplexing, routing, protection and timing functions. These functions are technology dependent, for example, PDH, SDH, OTN or Ethernet. Each function provides a specific means to transfer client information such as voice, video and data. The transfer can be, for example, circuit-switched (voice) or connectionless (packets).

The complexity of the network elements has increased during the evolution of the digital telecommunication network. The initial PDH network contained multiplexers; the first generation SDH equipment were the *Add-Drop Multiplexers* (ADM) and the *Digital Cross-Connects* (DXC), all designed to transport voice signals. The next generation network elements are *Multi-Service Transport Platforms* (MSTP), capable of also transporting data signals over the SDH/SONET network. Figure 2.1 shows the evolution through time and in equipment complexity.

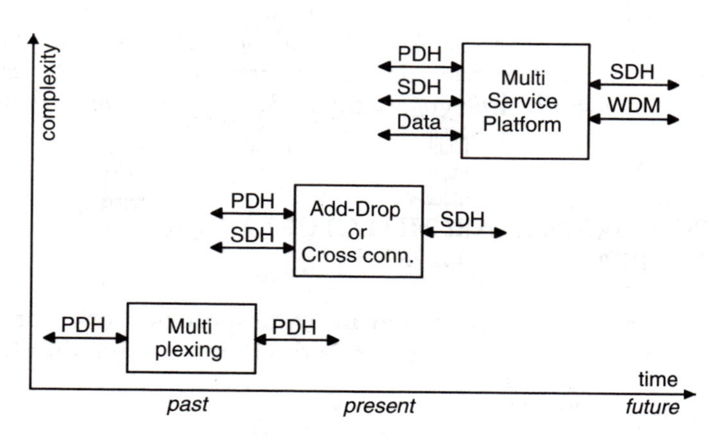

Figure 2.1 Network evolution.

There are several types of MSTPs distinguishable:

- *Multi-Service Switching Platforms* (MSSP) for deployment in high-capacity core networks. At the line side they have interfaces such as STM-64/OC-192 and STM-256/OC-768 for the high bandwidth metro aggregation. There are also systems that do have integrated DWDM interfaces. At the tributary side the interfaces are STM16/OC-48 and also high speed data, e.g. 10 GbE. This multi-service functionality allows service providers to support current TDM services and carry the benefits of next generation services (such as Ethernet) into the central office while still utilizing their existing SONET or SDH infrastructure.
- *Multi-Service Provisioning Platforms* (MSPP) for metro edge aggregation networks. The line interfaces are STM-4/OC12 and STM-16/OC-48. At the tributary side they have STM-1/OC-3 and STM-4/OC-12 interfaces as well as data service signals like Gigabit Ethernet, Fibre Channel, FICON, ESCON, etc.
- *Multi-Service Access Platforms* (MSAP) deployed in edge networks. The line interfaces are STM-1/OC3 and STM-4/OC12. At the tributary side they can accommodate PDH signals, STM-0/OC1 and STM-1/OC3 signals, and data service signals like (Fast) Ethernet,

Figure 2.2 shows where the platforms are located in a typical transport network. These platforms allow service providers to build high-bandwidth systems that integrate core and edge networks and

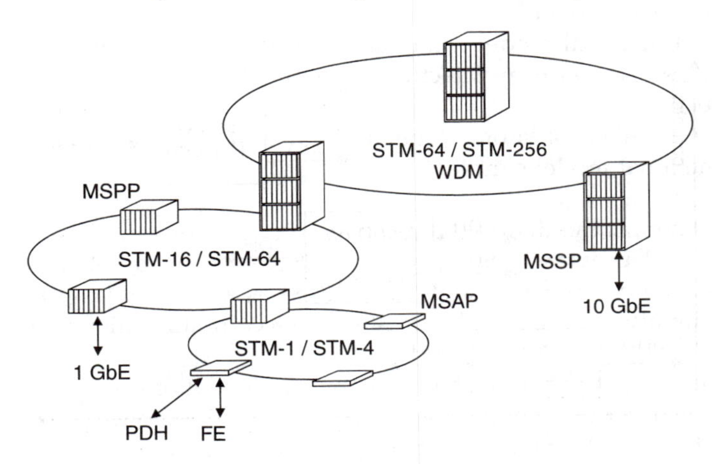

Figure 2.2 Typical transport network.

handle voice, data, video and other services, while reducing deployment and operating costs. They deliver the traffic management and connectivity capabilities needed to implement virtual private networks (VPN), bandwidth provisioning between dissimilar networks, and more. With these systems the providers can provide *Local Area Networks* (LAN), *Metro Area Networks* (MAN), *Wide Area Networks* (WAN) and *Storage Area Networks* (SAN) very efficiently.

An appropriate network model is essential to be able to describe this complex network. This model has to use well-defined functional entities to be able to design and manage the actual network as accurately as possible. Similar to an electronic network or circuit, a transport network can be described by defining the associations between points in that network. To keep the description simple, the transport network model is based on the concept of using separate layers and partitions within each layer. In this way a high degree of recursion can be provided.

2.2 FUNCTIONAL MODEL REQUIREMENTS

The following requirements were set during the definition phase of the concept of using atomic functions to model the transmission network:

- The resulting functional model shall present the *functional behavior* of the implementation and *not* the implementation itself. This ensures that many different implementations will fit the same functional model.
- The functional model shall *not* describe the underlying hardware and/or software architecture because these are implementation specific.
- The number of atomic functions shall be limited to keep the functional model simple.

As a bonus, the definition of the atomic functions in the functional model will provide a structured and well-organized set of requirements. The functional model can be used as a *common language* at all levels involved in the deployment of a telecommunications network:

- in telecommunication standards recommendations;
- in the description of the layered network by the service provider for both network management purposes and for the physical deployment of equipment and interconnecting fibers;

- in the requests for information, such as requests for quote and/or service level agreement documentation exchanged between service providers and equipment manufacturers;
- in the sales documentation of equipment vendors;
- in the (internal) equipment architecture and specification of manufacturers;
- in the equipment development requirements;
- in the telecommunication device maker specifications;
- in the device development specifications.

2.3 FUNCTIONAL MODEL BASIC STRUCTURE

In a functional model a subdivision can be made:

- architectural components;
- topological components.

2.3.1 Architectural components

A thorough analysis of the existing transport networks was performed to identify a set of generic functions that could be used to define the model. The result of the analysis is a set of atomic functions that will provide a means to describe the functionality of a transport network in an abstract way by using only a small number of architectural components. These architectural components are the atomic functions that are defined either by the task they perform in terms of the information they process or by their description of relationships between other adjacent architectural components. In general, the atomic functions that are currently defined in the standards will process the information that is presented at one or more of their inputs and then present the processed information at one or more of their outputs. Each component or function is defined and characterized by the information process between its inputs and outputs. The architectural components can be associated with each other following the connection rules to construct a network element. A transport network can be built using the models for the network elements. In the transport network architecture, reference points can be identified that are the result of the binding of the inputs and outputs of processing functions and transport entities.

For each generic function a specific symbol has been defined as well as the connection rules. These functions and their symbols will be described next and are also illustrated in Figures 2.3 to 2.6.

- *Input or Output.* This symbol is used to indicate the direction of the flow of information and is part of an atomic function.

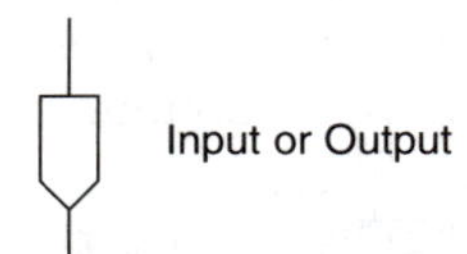

Figure 2.3 Input/output symbol.

- *Connection atomic function.* This symbol represents the connectivity available in a network element and in a network. Connections in a connection function are made between an input and one or more outputs and can also be removed. A connection can be provisioned by the Element Management System (EMS).

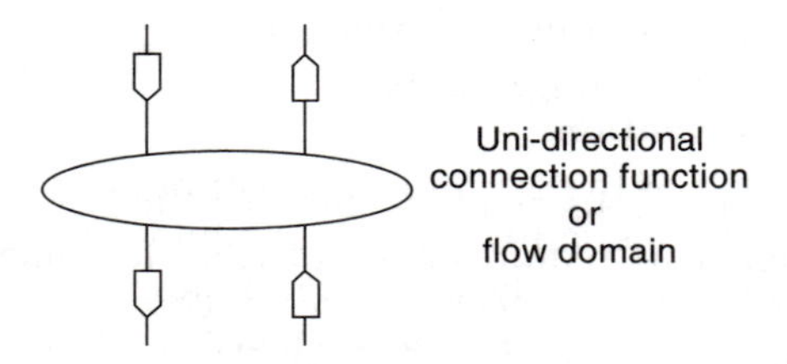

Figure 2.4 Connection or flow domain atomic function.

The connection function is defined for a connection-oriented network. The equivalent function in a connectionless network is referred to as a flow domain because the information is transferred in flows instead of over connections.

- *Adaptation atomic function.* This symbol represents the adaptation of information structure present in the client layer network to a structure that can be transported in the server layer network.

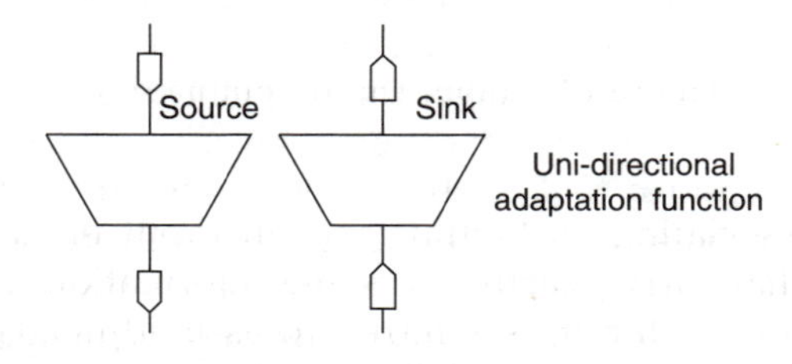

Figure 2.5 Adaptation atomic function.

- *Trail termination atomic function*. This symbol represents the start and end points, or Source and Sink, of a trail through the transport layer network.

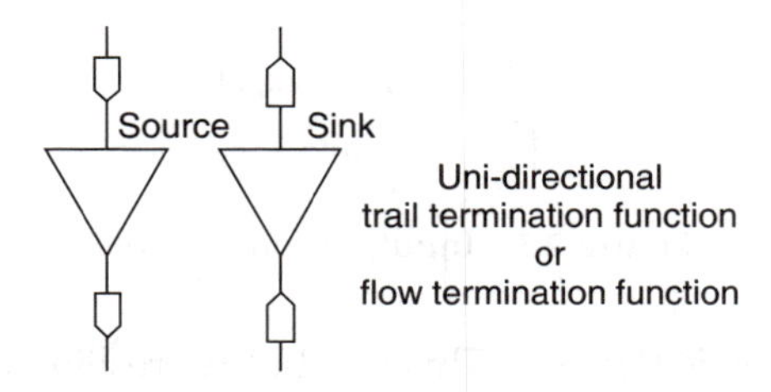

Figure 2.6 Trail or flow termination atomic function.

The trail termination function is defined for a connection-oriented network. The equivalent function in a connectionless network is referred to as a flow termination function.

Diagrammatic conventions

Of course, these functions shall be combined to construct a complete functional model. This requires a set of rules for interconnecting the atomic functions. These connection rules or conventions are described below and depicted in Figures 2.7 to 2.9:

- *Pairing*. Associating an Input and an Output of the same atomic function that carry exactly the same information to form a bi-directional port is referred to as pairing.

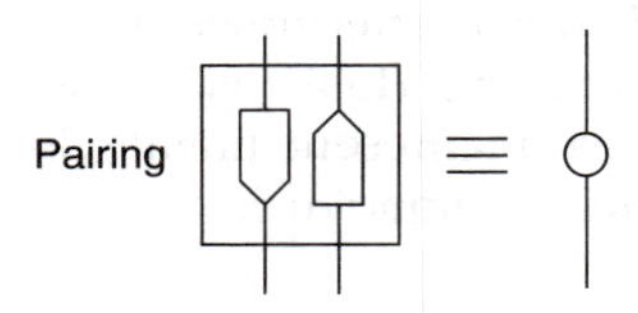

Figure 2.7 Input and output pairing.

- *Binding*. Associating an Output and an Input of different atomic functions that carry exactly the same information, i.e. connecting an Output to an Input, is referred to as binding. The connection itself is referred to as a *uni-directional reference point*. In general, connectionless transport is uni-directional thus this binding is used as the generic reference point in connectionless networks.

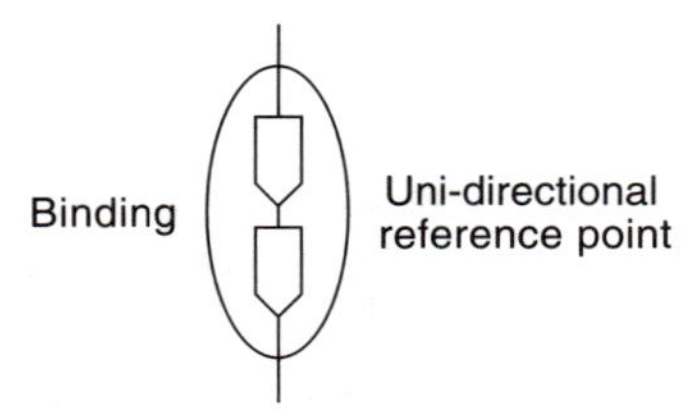

Figure 2.8 Input and output binding.

- *Reference Point*. More exactly a *bi-directional reference point*, this is a combination of pairing and binding of Inputs and Outputs of atomic functions. It is the reference point commonly used in connection oriented networks. However, it is also used as a shorthand notation for two co-located Flow Points in opposite directions.

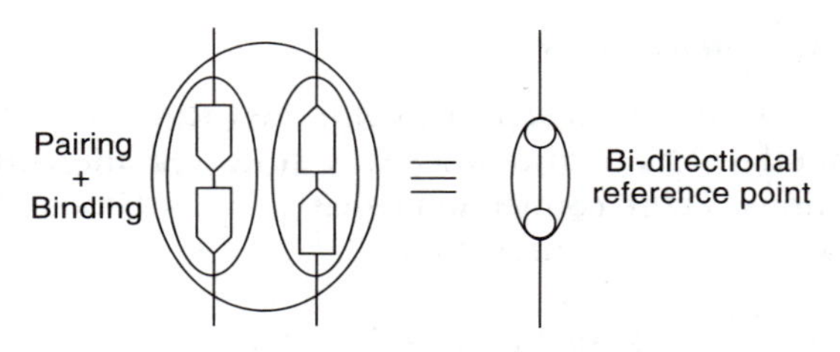

Figure 2.9 Pairing and binding.

Most of the time the trails through a connection oriented network are bi-directional and to keep the functional models simple all functions in the model can be depicted as a bi-directional symbol. Figures 2.10 to 2.12 show the bi-directional symbols of the connection, adaptation and termination functions.

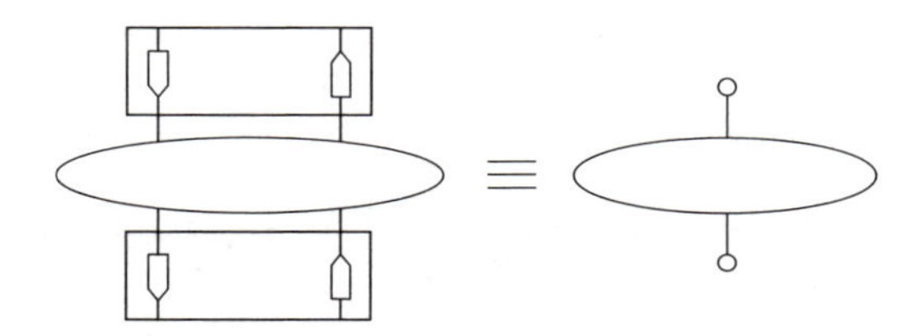

Figure 2.10 Bi-directional connection function.

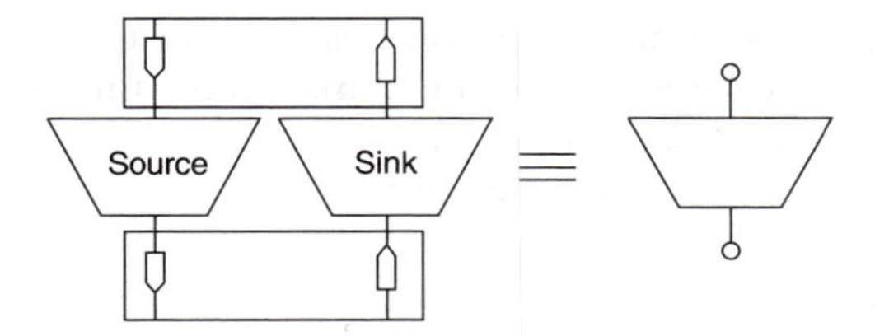

Figure 2.11 Bi-directional adaptation function.

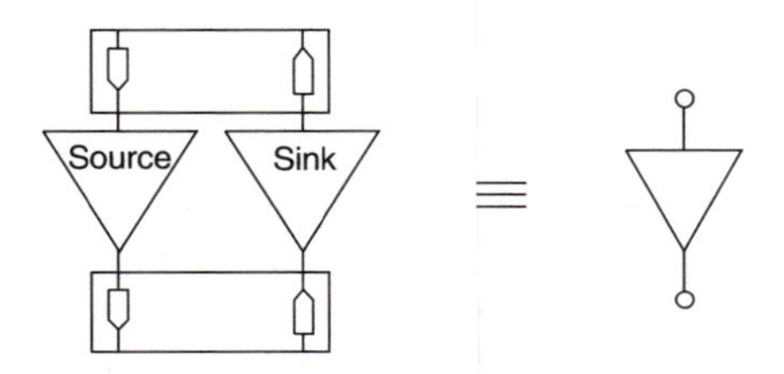

Figure 2.12 Bi-directional trail termination function.

Relation of functions in a model

The relation between the atomic functions and the layered network is illustrated in Figure 2.13. This figure also shows the successive order of the atomic functions in a functional model.

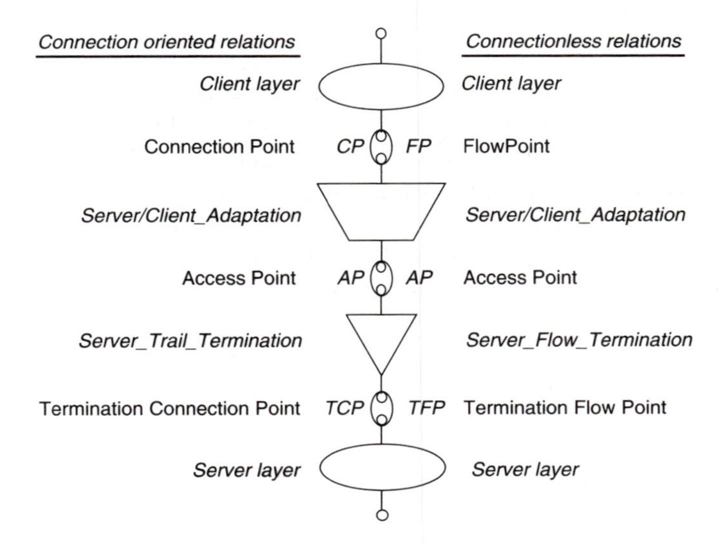

Figure 2.13 Functional model logical order.

Since, in general, there is a one-to-one relation between an adaptation function, the access point and the associated trail or flow termination function, these three functional components can be represented by a single symbol, i.e. a compound function, as shown in Figure 2.14.

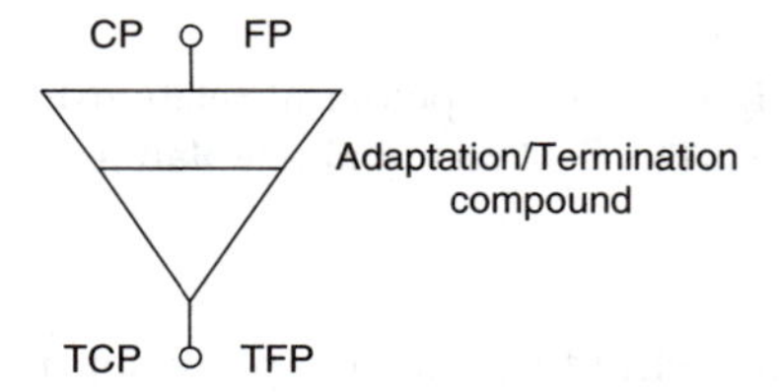

Figure 2.14 Compound function.

2.3.2 *Topological components*

The topological components can be used to provide the most abstract description of a network in terms of the topological relationships between sets of similar reference points. Four topological components have been distinguished as follows:

- the *layer network*;
- the *sub-network*;
- the *link*; and
- the *access group*.

By using only these components it is possible to describe completely the logical topology of a layer network. The topological components and their relationships are illustrated in Figures 2.15 and 2.16.

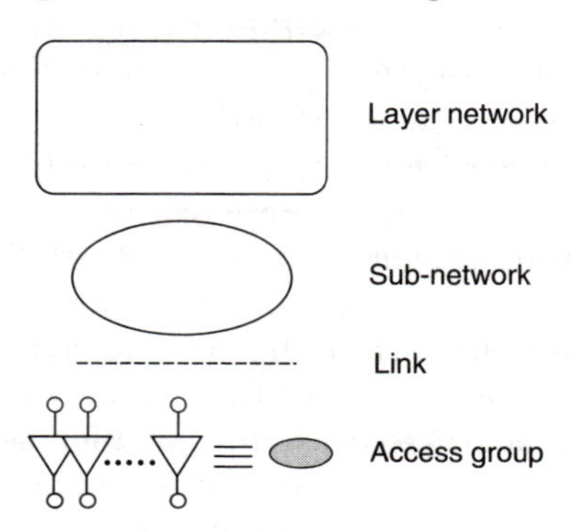

Figure 2.15 Topological components conventions.

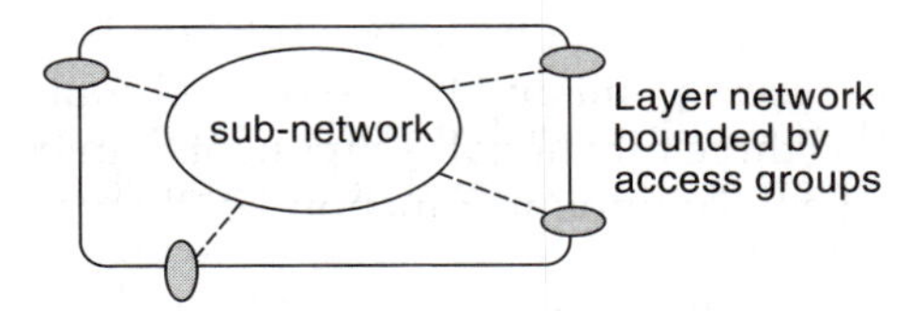

Figure 2.16 Topological relationships.

Layer network

A layer network is defined by the complete set of *access groups* of the same type that can be associated for the purpose of transferring information. The information that is transferred is characteristic of a specific layer network and is termed characteristic information. The information is transferred over a trail between two or more trail termination points in the layer network. The association of the trail terminations to form a trail in a layer network may be provisioned, i.e. made and broken, by a layer network management process. Changing the associations between trail terminations is equivalent to changing the layer network connectivity. For each type of trail termination there exists a separate and logically distinct layer network. Access groups and sub-networks are the components used to describe the structure of layer networks. The *links* between the access groups and sub-networks describe the topology of a layer network.

Sub-network

A sub-network will only exist within a single layer network. It consists of the set of ports that are available for the purpose of transferring the layer network characteristic information. The associations between two or more ports at the edge of a sub-network may be provisioned by a layer network management process and will change the sub-network connectivity. At the moment a sub-network connection is provisioned, the related reference points are also created. These reference points are created when the ports are bound to the input and output of the sub-network connection. It is allowed to divide sub-networks into smaller sub-networks; these smaller sub-networks are then interconnected by links. A connection matrix is a special case of a sub-network that cannot be further partitioned.

Link

A link is defined as a subset of the ports at the edge of a particular sub-network or of an access group that are associated with a corresponding subset of the ports at the edge of another sub-network or access group for the purpose of transferring the layer network characteristic information. A link represents the topological relationship and the available transport capacity between a pair of sub-networks, or a sub-network and an access group or a pair of access groups. Multiple links may exist between any given sub-network and an access group or any pair of sub-networks or access groups within a single layer network.

Generally, links are provisioned and maintained based on the existence of the server layer network. However, they are not necessarily limited to being provided by a server trail; they can also be provided by client layer network connections, e.g. by using inverse multiplexing.

Access group

An access group is defined as a group of co-located trail termination functions connected to the same sub-network or to the same link.

2.4 FUNCTIONAL MODEL DETAILED STRUCTURE

In this section a more detailed description is provided of the basic structures; the following distinctions are made in the detailed structure:

- transport entities;
- transport processing functions;
- reference points.

2.4.1 Transport entities

The transport entities provide the transparent transfer of information between two or more layer network reference points. The information available at the output is exactly the same information presented at the input unless it is affected by degradation, e.g. bit errors, of the transfer process.

Two basic transport entities can be distinguished based on the capability to monitor the integrity of the transferred information. These are termed

- *Connections*. Connections can be distinguished further by the topological component to which they belong:

  - *network connections*;
  - *sub-network connections*; and
  - *link connections*.

- *Trails*.

Figure 2.17 shows these transport entities.

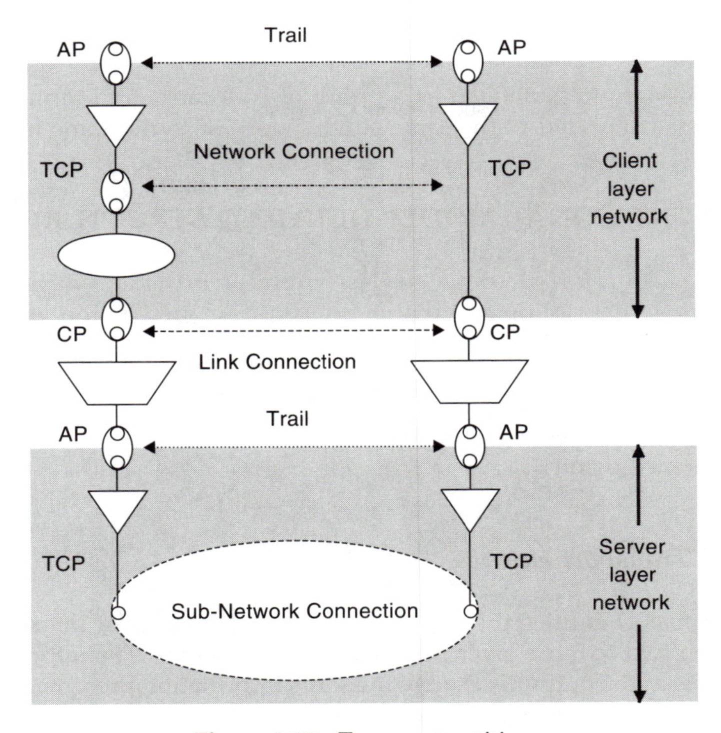

Figure 2.17 Transport entities.

Connections

Link connection

A link connection is defined as the capability to transfer information transparently across a link. It represents the fixed relation between the input and output ports at each end of the link and the associated trail through the network. In other words: a link connection represents a pair of adaptation functions and a trail in the server layer network.

The port at the input to a uni-directional link connection is equivalent to the input of an adaptation source function and the port at the output of a uni-directional link connection is equivalent to the output of an adaptation sink function. One uni-directional link connection and the associated uni-directional link connection in the opposite direction together with their associated ports and adaptation sink and source functions may be paired to provide a bi-directional transfer of information. Figure 2.18 shows the network associations in a functional model.

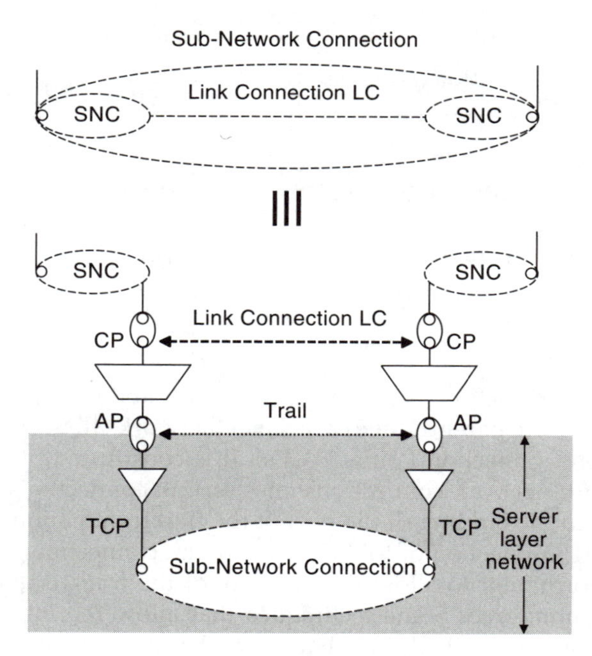

Figure 2.18 Sub-network equivalence.

Sub-network connection

A sub-network connection is defined as the capability of transferring information transparently across a sub-network. It is delimited by ports at the boundary of the sub-network and represents the association between these ports.

Sub-network connections are generally constituted by a concatenation of sub-network connections and link connections. A matrix connection is a special case of the sub-network connection and consists of a single (indivisible) sub-network connection. This is depicted in Figure 2.19.

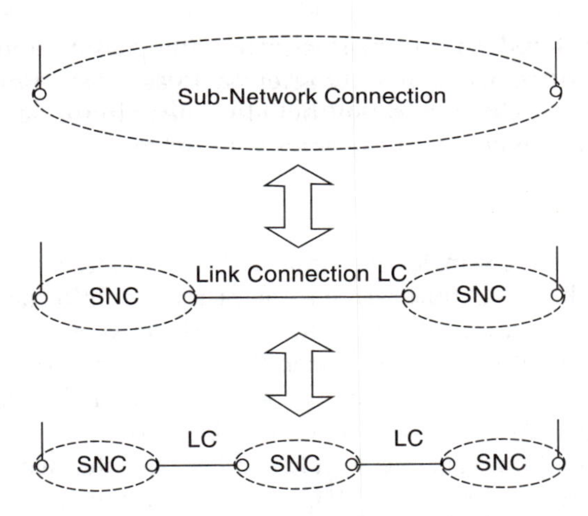

Figure 2.19 Sub-network constitution.

Network connection

A network connection is defined as the capability of transferring information transparently across a layer network. It is delimited by Termination Connection Points (TCPs). It is constituted by a concatenation of sub-network connections and/or link connections. A TCP is equivalent to the binding of the port of the trail termination to either a sub-network connection or to the port of a link connection. There is no explicit information to allow the integrity of the transferred information to be monitored. Some techniques that allow the integrity to be monitored are described in Chapter 8. Figure 2.20 illustrates the concatenation of a network connection. If part of a network connection

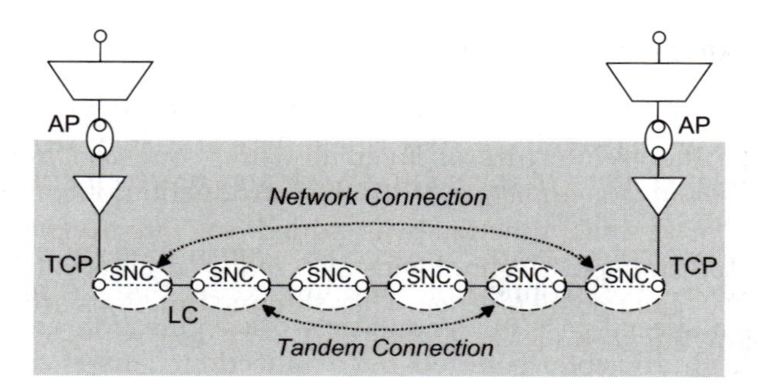

Figure 2.20 Network connection concatenation.

has to be monitored, for example, because it is passing another service providers domain, it is generally referred to as a tandem connection.

Chapter 7 provides a more detailed description of connection functions and also contains example applications.

Trails

A trail represents the transfer of characteristic information (CI) of the client layer. To enable this transfer, the characteristic information will be adapted for the transport through the network between the access points (AP) and it is monitored for the performance of the transport. Two access points delimit a trail, one at each end of the trail. A trail is defined as the association between the access points at each end of the trail. To provision a trail, an association has to be established between the two trail terminations and a network connection. Figure 2.21 shows the association.

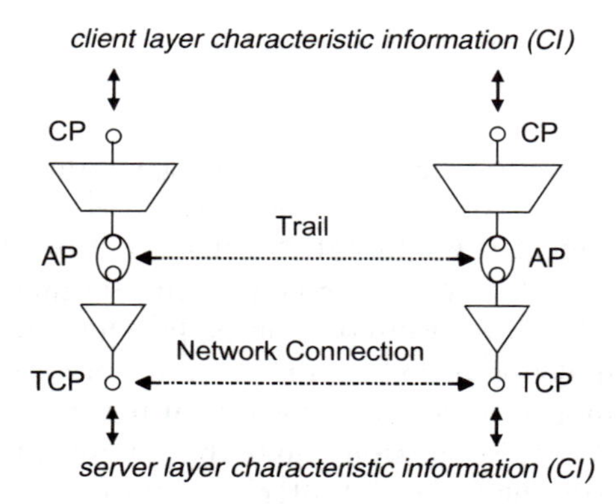

Figure 2.21 Trail association.

2.4.2 *Transport processing functions*

Initially, there were two distinct generic processing functions defined to describe the architecture of layer networks: an *adaptation function* and a *trail termination function*. With the introduction of concatenation one more processing function was added: an *interworking function*. Packet based transport technologies also required an additional function: a *traffic conditioning function*. These atomic functions are described in the following sections.

Adaptation function

- *Adaptation source function.* A transport processing function that adapts the client layer network characteristic information into a form suitable for transport over a trail in the server layer network.
- *Adaptation sink function.* A transport processing function that recovers the characteristic information of the client layer network from the server layer network trail information.
- *Bi-directional adaptation function.* A transport processing function that consists of a co-located adaptation source and sink pair.
- *Adaptation function I/O.* Several configurations are possible:

  - In the adaptation source function, one or more client layer network characteristic information streams are adapted into a single adapted information stream suitable for transport over a trail in the server layer network. The adaptation sink function provides the inverse functionality.

    This configuration is commonly used to represent the multiplexing of several client signals into a single server signal. The client signals are not necessarily originating from the same layer network.
  - In the adaptation source function a single client layer network characteristic information stream is split over several outputs and in the adaptation sink function the client information stream is reconstructed from signals present at the inputs. This configuration is used to represent inverse multiplexing.
  - In the adaptation source and sink functions, the client layer may be either connection oriented or connectionless and the server layer may also be either connection oriented or connectionless, see Figure 2.22 for the possible configurations.

- *Adaptation processes.* The following processes can be present in an adaptation function either as a single process or in combination with others:

  - scrambling, coding—descrambling, decoding;
  - rate adaptation, frequency justification;
  - transfer protection switch status, synchronization status;
  - alignment;
  - payload type identification;
  - Multiplexing.

- *Adaptation function models.*

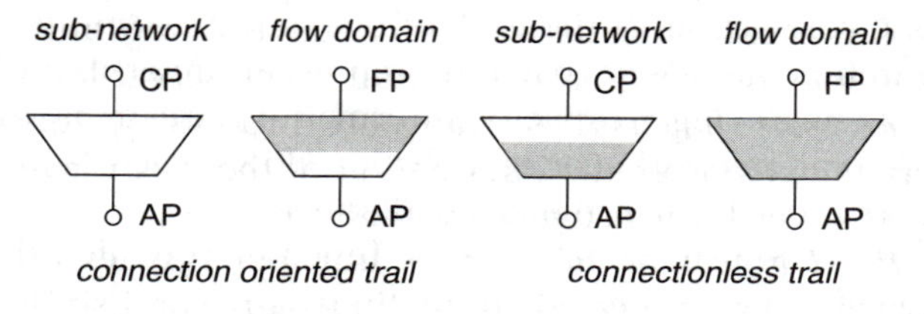

Figure 2.22 Adaptation function in possible configurations.

Chapter 5 provides a more detailed description of adaptation functions and also contains example applications.

Trail termination function

The equivalent flow termination function is generally uni-directional due to the nature of the connectionless network.

- *Trail/flow termination source function.* A transport processing function that takes the adapted client layer characteristic information presented at its input and adds information to provide monitoring of the trail. The resulting characteristic information of the server layer network is presented at its output. A trail termination source function can operate without an input from a client layer network.
- *Trail/flow termination sink function.* A transport processing function that accepts the characteristic information of the server layer network at its input, removes the information related to the performance monitoring of the trail and presents the remaining client

layer network information at its output. A trail termination sink function can operate without an output to a client layer network.

- *Bi-directional trail termination function.* A transport processing function that consists of a pair of co-located trail termination source and sink functions.
- *Trail/flow termination function I/O.* The trail/flow termination source function always has one input. There can be one or more outputs present in a single trail/flow termination function. The single adapted client layer network characteristic information input stream is distributed over one or more network connections or flows in the server layer. The one input to one output configuration is used most generally. It represents the addition of the Trail Overhead to the adapted information to be transported via a single network connection or flow. The one input to multiple output configuration could be used to represent inverse multiplexing where a single high capacity information stream is distributed over several network connections or flows each with a lower transport capacity then the original stream.

  The trail/flow termination sink function provides the inverse functionality and has one or more inputs and single output.
- *Trail/flow termination processes.* The following processes may be present in an adaptation function either as a single process or in combination with others:

  - scrambling – descrambling;
  - error detecting code generation – checking;
  - trail identification – connectivity check;
  - near-end and far-end performance monitoring.

- *Trail/flow termination function models.*

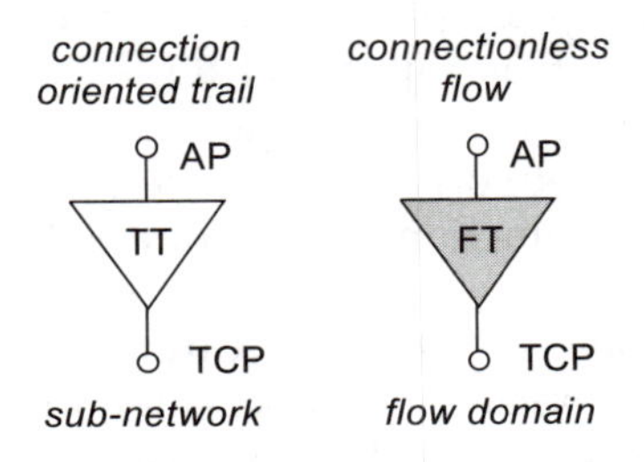

Figure 2.23 Trail and flow termination function.

Chapter 6 provides a more detailed description of trail termination functions and also contains example applications.

Interworking function

- *Interworking source function, contiguous to virtual.* A transport processing function that provides interworking between layer networks with the same characteristic information but with different transport structures. The function adapts a contiguous concatenated structure into a virtual concatenated structure and vice versa. It enables the transport of the client payload over multiple trails in a layer network that does not support the original contiguous concatenated structures.
- *Interworking sink function, virtual to contiguous.* The transport processing function that performs the inverse of the interworking source function, i.e. it converts the layer network trail information that uses a virtual concatenated structure into the characteristic information of the layer network that uses a contiguous concatenated structure.
- *Interworking function I/O.* The interworking source function has one input and multiple outputs and the interworking sink function has multiple inputs and a single output. The transport capability of the contiguous concatenated structure and the virtual concatenated structure is exactly the same.
- *Interworking processes.* The following processes are present in the interworking function:

  - trail overhead distribution and reconstruction;
  - framing;
  - alignment.

- *Interworking function model.*

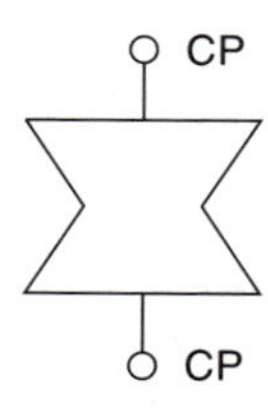

Figure 2.24 Interworking function.

Chapter 10 (Section 10.4) provides a more detailed description of interworking functions and contains an example application.

Traffic conditioning

- *Traffic conditioning function.* A flow processing function with the objective to determine the conformance of incoming Ethernet frames.
- *Traffic conditioning function I/O.* This function is uni-directional and has only a single input and a single output.
- *Traffic conditioning processes.* The complete Ethernet traffic conditioning process is still under study by ITU-T Studygroup 13 and so what follows is a preliminary definition. The level of conformance is expressed as one of three colors: Green, Yellow or Red. For a sequence of ingress Ethernet frames, $\{tj, lj\}j \geq 0$, with arrival times tj and lengths lj, the color assigned to each frame during traffic conditioning is defined by a Bandwidth Profile algorithm. Compliance to a Bandwidth Profile is described by four parameters:

  (1) Committed Information Rate (CIR) expressed as bits per second. CIR must be ≥ 0.
  (2) Committed Burst Size (CBS) expressed as bytes. When $\text{CIR} > 0$, CBS must be $\geq$ Maximum Ethernet frame allowed to enter the network.
  (3) Excess Information Rate (EIR) expressed as bits per second. EIR must be ≥ 0.
  (4) Excess Burst Size (EBS) expressed as bytes. When $\text{EIR} > 0$, EBS must be $\geq$ Maximum Ethernet frame allowed to enter the network.

  Two additional parameters are used to determine the behavior of the Bandwidth Profile algorithm. The algorithm is said to be in color aware mode when each incoming Ethernet Frame already has a level of conformance color associated with it and that color is taken into account in determining the level of conformance to the bandwidth profile parameters. The Bandwidth Profile algorithm is said to be in color blind mode when level of conformance color (if any) already associated with each incoming Ethernet Frame is ignored in determining the level of conformance. Color blind mode support is required at the UNI. Color aware mode is optional at the UNI.

- Coupling Flag (CF) must have only one of two possible values, 0 or 1.
- Color Mode (CM) must have only one of two possible values, 'blind-blind' and 'aware-aware'.

• *Traffic conditioning model.*

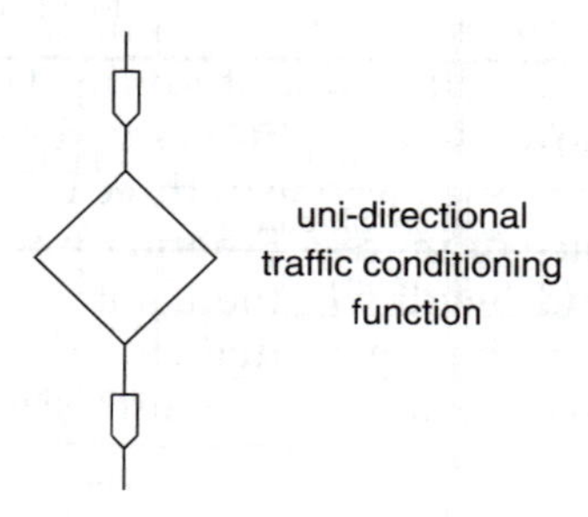

Figure 2.25 Traffic conditioning model.

2.4.3 *Reference points*

Reference points are defined as the binding between inputs and outputs of transport processing functions and/or transport entities. The allowable bindings for uni-directional and bi-directional architectures and the resulting specific types of reference points are shown in Table 2.1 and are also illustrated in Figure 2.26. Note that these reference points are all concerned with the transport of client and server layer characteristic information over the (termination) connection points ((T)CP) and server layer adapted information over the access points (AP). In Ethernet, the characteristic information consists of flows and so in general the (T)CP are referred to as (termination) flow points ((T)FP).

Note that in Table 2.1 the term 'Sub-Network Connection' is used for the Sub-Network Connection function.

The equivalent Ethernet model is depicted in Figure 2.27.

Table 2.1 Allowable bindings and resulting reference points.

Architectural components				Reference point	
Adaptation	So output	So input	Trail Termination Flow Termination	AP	uni-dir
	Sk input	Sk output			uni-dir
	So/Sk pair	So/Sk pair			bi-dir
Trail Termination Flow Termination	So output	input	Sub-Network Connection Flow Domain	TCP TFP	uni-dir
	Sk input	output			uni-dir
	So/Sk pair	bi-dir			bi-dir
Sub-Network Connection Flow Domain	input	So output	Adaptation	CP	uni-dir
	output	Sk input			uni-dir
	bi-dir	So/Sk pair			bi-dir
Trail Termination Flow Termination	So output	input	Link Connection Link Flow	TCP TFP	uni-dir
	Sk input	output			uni-dir
	So/Sk pair	bi-dir			bi-dir
Trail Termination Flow Termination	So output	Sk input	Trail Termination Flow Termination	TCP TFP	uni-dir
	Sk input	So output			uni-dir
	So/Sk pair	bi-dir			bi-dir
Trail Termination Flow Termination	Sk input	Sk output	Adaptation	TCP TFP	uni-dir
	So output	So input			uni-dir
	bi-dir	So/Sk pair			bi-dir
Link Connection Link Flow	output	input	Sub-Network Connection Flow Domain	CP FP	uni-dir
	input	output			uni-dir
	bi-dir	bi-dir			bi-dir
Adaptation	So input	Sk output	Adaptation	CP FP	uni-dir
	Sk output	So input			uni-dir
	So/Sk pair	So/Sk pair			bi-dir

AP Access Point
CP Connection Point
TCP Termination Connection Point
FP Flow Point
TFP Termination Flow Point

bi-dir bi-directional
uni-dir uni-directional
So/Sk Source/Sink

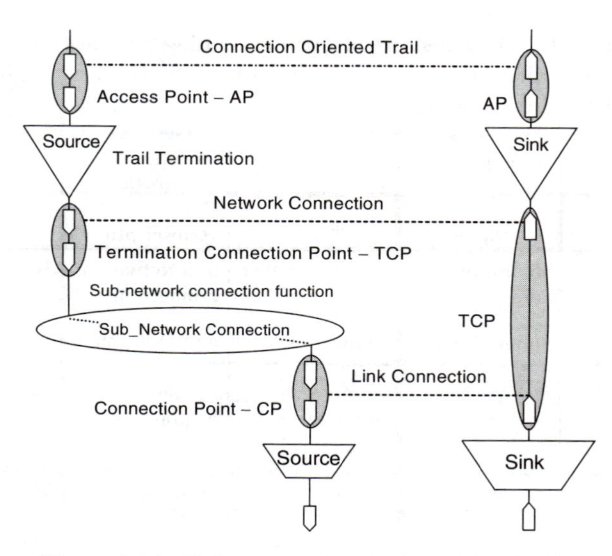

Figure 2.26 Reference point types and bindings.

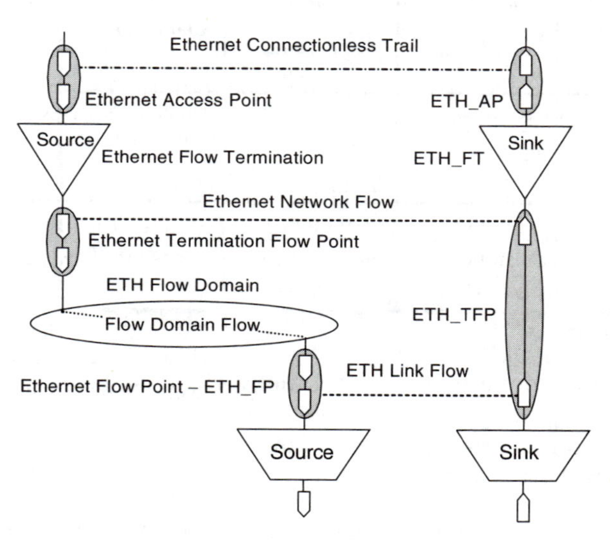

Figure 2.27 Ethernet reference point types and bindings.

In addition to the transport reference points the following reference points are also present (Table 2.2), but may not always be shown in a functional model:

Table 2.2 Additional reference points.

Reference point	Name	Description
MP	Management Point	Passes management information _MI, e.g. function provisioning, performance primitives, probable causes.
TP	Timing Point	Passes synchronization information _TI, e.g. clock, framestart.
RP	Remote Point	Passes remote information _RI between associated sink and source functions, i.e. RDI, REI.
PP	Replication Point	Replicates Ethernet characteristic information ETH_CI from the Ethernet adaptation source function to the Ethernet adaptation sink function.

2.4.4 *Components comparison*

The description of the architectural components in the previous sub-sections is mainly based on ITU-T Rec. G.805 (2000). This architecture is connection oriented. Occasionally, for the connectionless architecture components, a reference was made to ITU-T Rec. G.809 (2003). This sub-section provides a comparison of the components used in both architectures in Table 2.3.

It should be noted that even though the terms *access group* and *access point* appear to be synonymous in both architectures, their definitions are different:

- The access group in a connection oriented layer network is defined in terms of trail terminations, sub-networks and links; in a connectionless layer network it is defined in terms of flow terminations, flow domains and flow point pool links.
- The access point in a connection oriented layer network is the reference point between the adaptation and trail termination function; in a connectionless layer network it is the reference point between the adaptation and flow termination function. Furthermore, the connection oriented access point is associated with a trail and the connectionless access point is associated with a connectionless trail.

This allows for a common description of a layer network for both connection oriented and connectionless cases.

Table 2.3 Components of connection oriented and connectionless architectures compared.

Connection oriented component	Connectionless component
Access group	Access group
Access point	Access point
Adaptation	Adaptation
Adaptation sink	Adaptation sink
Adaptation source	Adaptation source
Adapted information	Adapted information
Architectural component	Architectural component
Characteristic information	Characteristic information
Client/server relationship	Client/server relationship
Uni-directional connection	**Flow**
Uni-directional connection point	**Flow point**
Layer network	Layer network
Link	**Flow point pool link**
Link connection	**Link flow**
Network connection	**Network flow**
Sub-network	**Flow domain**
Sub-network connection	**Flow domain flow**
Uni-directional termination connection point	**Termination flow point**
Topological component	Topological component
Trail	**Connectionless trail**
Trail termination	**Flow termination**
Trail termination sink	**Flow termination sink**
Trail termination source	**Flow termination source**
Transport	Transport
Transport entity	Transport entity
Transport network	Transport network
Transport processing function	Transport processing function

It should be noted also that for the connection oriented layer network both uni-directional and bi-directional components are defined while for a connectionless layer network, due to its characteristics, only uni-directional components are defined. In Table 2.3, components specified in only one of the layer networks are indicated in **bold**.

2.5 CLIENT/SERVER RELATIONSHIP

The client/server relationship between adjacent layer networks is one where a link connection in the client layer network is supported by a trail in the server layer network as illustrated in Figure 2.28.

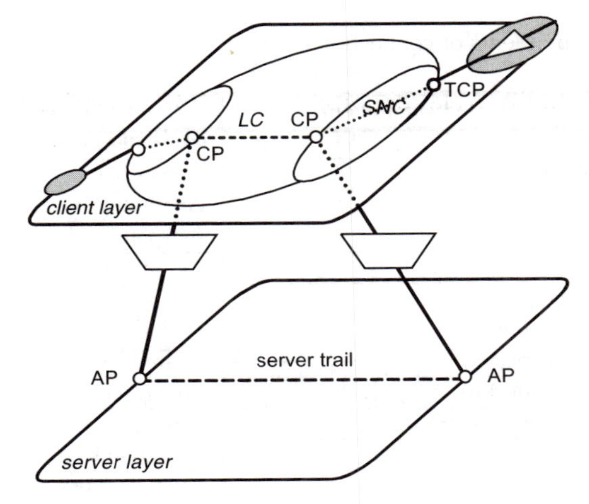

Figure 2.28 Layer network associations.

The concept of adaptation has been introduced to describe how the client layer network characteristic information is modified so that it can be transported over a trail in the server layer network. Therefore, from a transport network functional viewpoint the adaptation function falls between the layer networks. All the reference points belonging to a single layer network can be visualized as lying on a single plane.

In the example of the client layer network only a single boundary access group is shown containing only one trail termination function. The access group is bounded to the sub-network at its termination connection point (TCP). A TCP is connected to a connection point (CP) by a sub-network connection (SNC). The CPs are connected by link connections (LC). A (LC) represents the server layer trail between the access points (AP) at the boundary of the server layer network. (Note that this concept of contiguous layer boundaries in the transport network model is completely different from the layering concept used in the OSI protocol reference model described in ITU-T Rec. X.200 (1994)).

The client/server relationship can be:

- A one-to-one relationship; in this case a single client layer link connection is supported by a single server layer trail.
- A many-to-one relationship, generally referred to as multiplexing and described in the next section.
- A one-to-many relationship, also referred to as inverse multiplexing (see description in Section 2.5.2).

2.5.1 *Multiplexing*

In the many-to-one relationship, several link connections present in client layer networks are supported by a single server layer trail at the same time as shown in Figure 2.29.

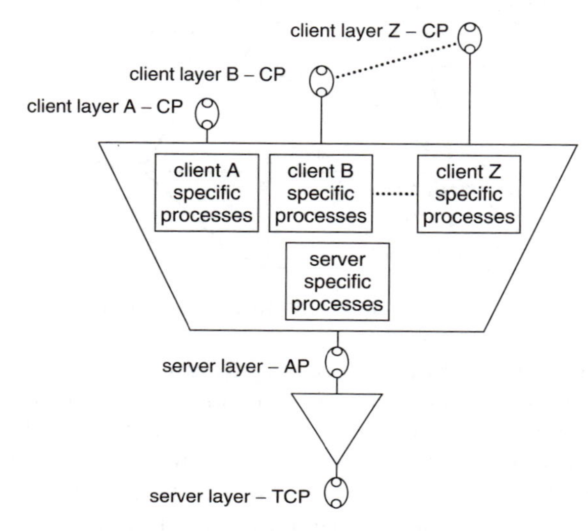

Figure 2.29 Multiplexing example.

Multiplexing techniques may be used to combine the client layer signals. The client signals may originate in different layer networks or in the same layer network, as shown in Figure 2.30.

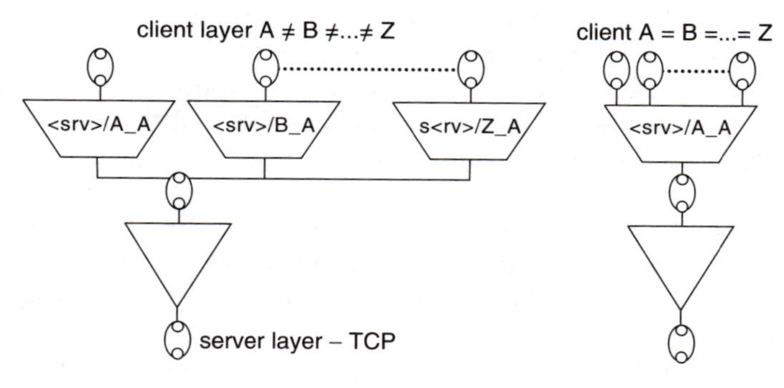

Figure 2.30 Alternative representation of multiplexing.

Examples of different client layers are the Sn and Sn–Xc transport layers and the *Data Communications Network* (DCN) layer. The Data Communications Channel (DCC) provides network connectivity for the DCN. Other general-purpose communications channels are the *Engineer Order Wire* (EOW) and the *User Channel* (USR).

The adaptation function will contain client specific processes to handle the client characteristic information present at the connection points and processes related to the server layer characteristic information.

2.5.2 *Inverse multiplexing*

In the one-to-many relationship, a single client layer link connection is supported by several server layer trails in parallel. Inverse multiplexing techniques (e.g. virtual concatenation, ATM inverse multiplexing) are used to distribute the client layer signal. The server signals may be of the same or of different types.

The generic functional model for inverse multiplexing is depicted in Figure 2.31. The support for inverse multiplexing is realized by introducing an inverse multiplexing sublayer. This layer consists of an inverse multiplexing trail termination function IM_TT and an inverse multiplexing adaptation function <*srv*>/IM_A where <*srv*> represents the server layer type. The IM_TT function will provide the trail performance monitoring of the composite signal. The <*srv*>/IM_A function provides the distribution and reassembly of the composite signal over and of the n individual server layer trails. The two inverse multiplex sublayer functions and the X server layer network termination functions <*srv*>_TT can be represented by the inverse multiplexing trail termination compound function IMc_TT.

The X server layer trails may follow different paths (diverse routing) through the network. The individual signals will have a different transport delay due to the differences in physical path length. The sink side function <*srv*>/IM_A_Sk has to compensate this delay before the reassembly of the composite signal can start. The difference between the shortest and the longest transport delay is generally referred to as differential delay. The maximum acceptable differential delay is application and/or implementation specific. Routing the individual signals along the same path in the network will minimize the differential delay.

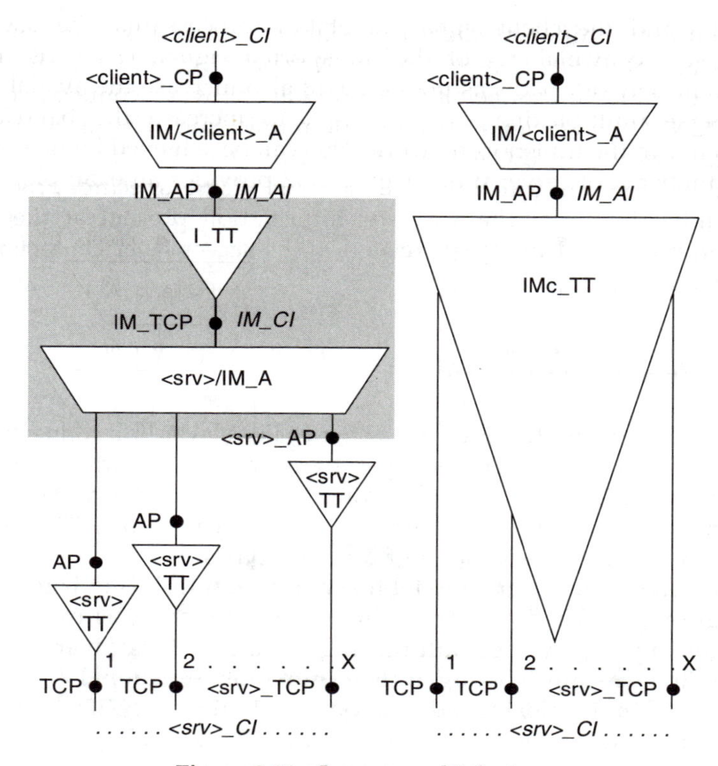

Figure 2.31 Inverse multiplexing.

The performance of the inverse multiplex sub-layer trail is determined by the performance of the X individual server layer trails monitored by the *<srv>*_TT_Sk functions and the performance monitored by the reassembly process present in the *<srv>*/IM_A_Sk function. At intermediate monitoring points, e.g. non-intrusive monitors, only the performance of the X individual server layer trails is available.

If a client signal has a fixed bandwidth, the number of server layer trails is fixed as well. Here the layer interworking, as defined in Section 2.6, may be applicable between the single server layer trail that supports the full client signal and the X individual server layer trails that support the inverse multiplexed client signal.

If a client signal has a variable bandwidth the number of server layer trails may vary too. The network operator, the client layer process or the protection switching process may change the number of server layer trails on request. Interworking between the inverse multiplexing process and the client signal adaptation process may be used to increase the availability of the transported signal. Using as many different network paths as possible will also increase the availability.

Inverse multiplexing may be used to increase the bandwidth efficiency in the transport network. This can be achieved by increasing the number X of server trails. Figure 2.32 provides an example.

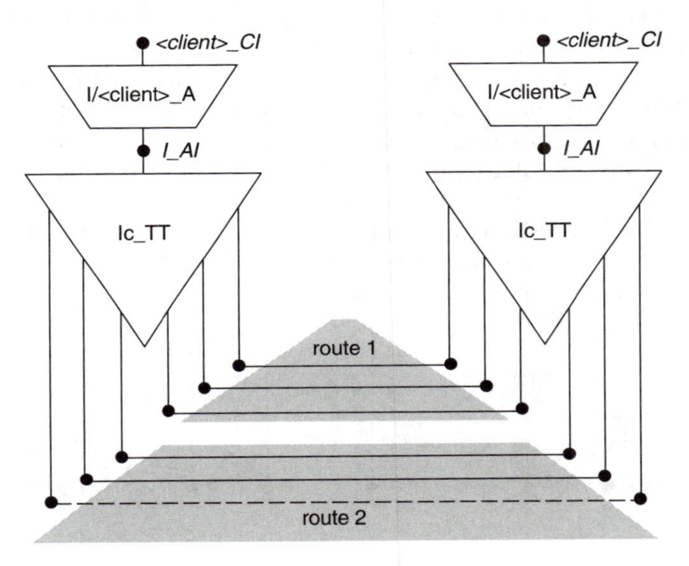

Figure 2.32 Bandwidth efficiency.

In this example, a client signal with variable bandwidth requires five server layer trails for its transport. Since only two separate routes are available, three server trails are routed via route 1 and two trails are routed via route 2. In this case, the survivable bandwidth is two trails out of five or 40%, and the non-survivable bandwidth is three trails out of five or 60%. By adding one extra trail and routing it via route 2 the survivable bandwidth will become 3/5 or 60% and the unprotected bandwidth 2/5 or 40%.

2.6 LAYER NETWORK INTERWORKING

The objective of layer network interworking is to provide an end-to-end trail between different types of layer network trail terminations.

This requires interworking of characteristic information as different layer networks have by definition different characteristic information. In general, the adapted information of different layer networks for the same client layer network is also different, although this is not necessarily the case. Therefore, layer networking may require the interworking of adapted information. The trail overhead of a layer network can be defined in terms of semantics and syntax. Provided that the same semantics exist in the two layer networks, the trail overhead can be interchanged by passing on the semantics from one layer network to the other in the appropriate syntax, as defined by the characteristic information. In other words, layer network interworking shall be transparent for the semantics of the trail overhead. If both layer networks have a different set of semantics, the layer network interworking is restricted to the common set of semantics. The layer network interworking function has to terminate (insert, supervise) the semantics that are not interchanged.

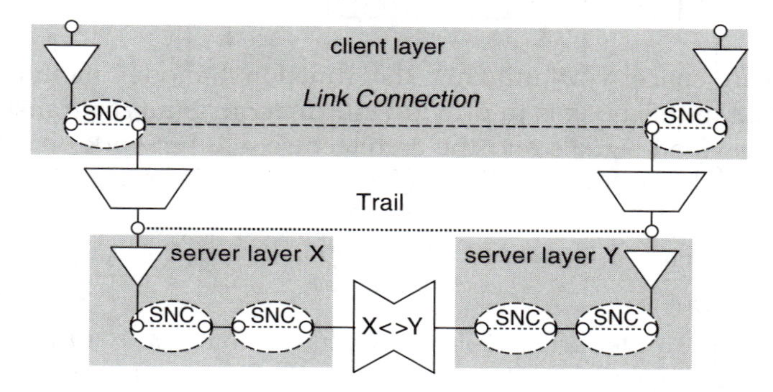

Figure 2.33 Layer network interworking.

Layer network interworking is accomplished through an interworking processing function as depicted in Figure 2.33. The interworking processing function supports an interworking link connection between two layer network connections. The interworking link connection is special in the sense that it is asymmetric, delimited by different types of ports. It is also special because in general it is only transparent for a specified set of client layers. An interworking link is a topological component that represents a bridge between two layer networks. The interworking link creates a 'super layer network', defined by the complete set of access groups that are interworking for a specified set of client layer networks.

- *Layer network interworking function, uni-directional.* A transport processing function that converts characteristic information of one layer network to the characteristic information of another layer network. The integrity of end-to-end performance and maintenance information is maintained. The function may be limited to a set of client layer networks.
- *Layer network interworking function, bi-directional.* A transport processing function that consists of a pair of co-located uni-directional service interworking functions, one for the interworking form layer network X to Y and the other for the interworking from layer network Y to X.

For an application of layer interworking see Chapter 10, Section 10.4.

2.7 LINKING THE FUNCTIONAL MODEL AND THE INFORMATION MODEL

The main concern of utilizing the functional model methodology described in this book is to provide information about the transformation of payload signals and the architecture of a network or network element. A basic functional model is shown again in Figure 2.34. As

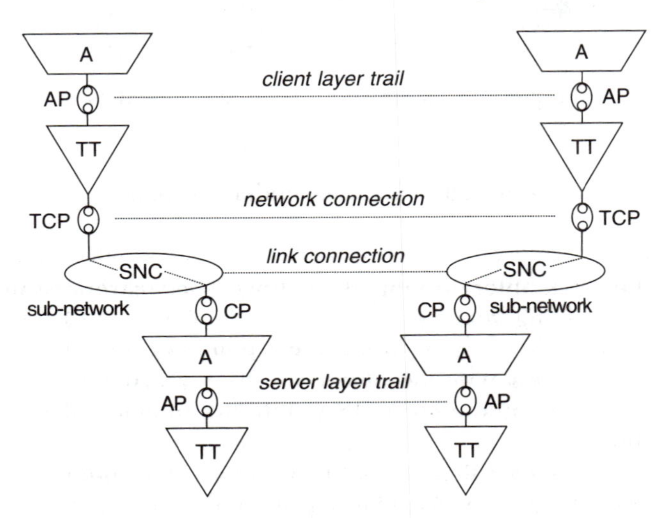

Figure 2.34 Basic functional model entities.

described in Section 2.4.2, the adaptation function provides the capability to transform the characteristic information of the client layer into that of the server layer. The trail termination function provides the verification of the correct transport of the information. The connection function provides the connectivity, i.e. sub-network connections (SNC) and link connections, within a single layer to enable the transport of this information between the sub-network connection points.

For the management and control of the connectivity in a sub-network, the information model methodology is used. ITU-T Recommendations M.3100 (1995) and G.8080 (2001) describe this methodology. Just for information, and to enable the comparison of both models, a basic information model is depicted in Figure 2.35.

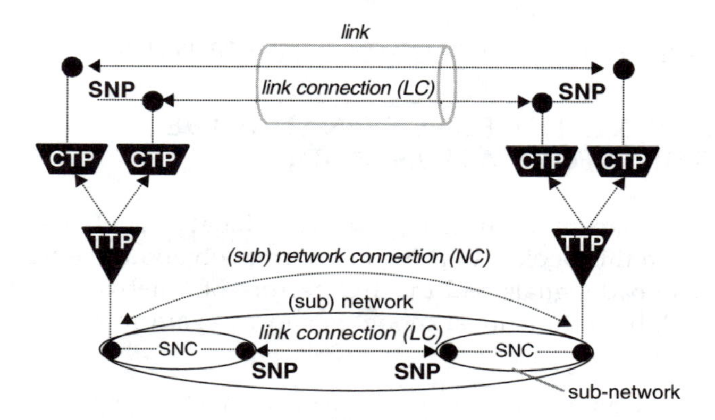

Figure 2.35 Basic information model entities.

The following terminology is info-model specific:

- *Connection Termination Point* (CTP): the client side of an adaptation function, it represents signal state at the input/output of an adaptation function.
- *Trail Termination Point* (TTP): the endpoints of a network connection that coincide with the ends of a trail in the functional model, it represents the signal state at the input/output of a trail termination function.
- *Sub-Network Points* (SNP): a pair of addressable connection points in the client network between which a link connection (LC) can be established and, as Figure 2.35 shows, this is also recursive.

Refer to ITU-T Rec. M.3100 (1995) and G.774 (2001) for complete definitions of CTP and TTP.

As demonstrated above, for the purpose of managing connections within a layer network, the underlying transport plane resources are represented by a number of entities in the control plane. Figure 2.36 illustrates the relationship between the transport resources described in ITU-T Rec. G.805 (2000), the entities that represent these resources for the purposes of network management (as described in ITU-T Rec. M.3100 (1995)) and the view of the transport resources as seen by the control plane.

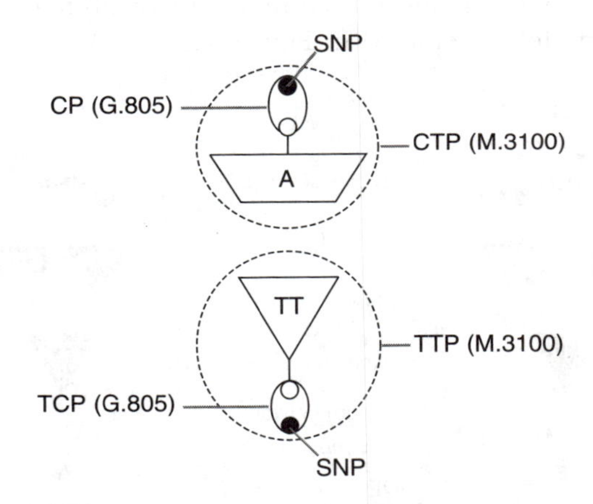

Figure 2.36 Relationship between model entities.

2.8 APPLICATION OF CONCEPTS TO NETWORK TOPOLOGIES AND STRUCTURES

2.8.1 PDH supported on SDH layer networks

Figure 2.37 shows an example in which a 2.048 kbit/s PDH signal is transported in an SDH network. Seven layer networks are involved in this transport:

(1) The E12 layer network terminates and transports the PDH intra-office section, i.e. the electrical signal on the interconnecting cable, as defined in ITU-T Rec. G.703 (1998).

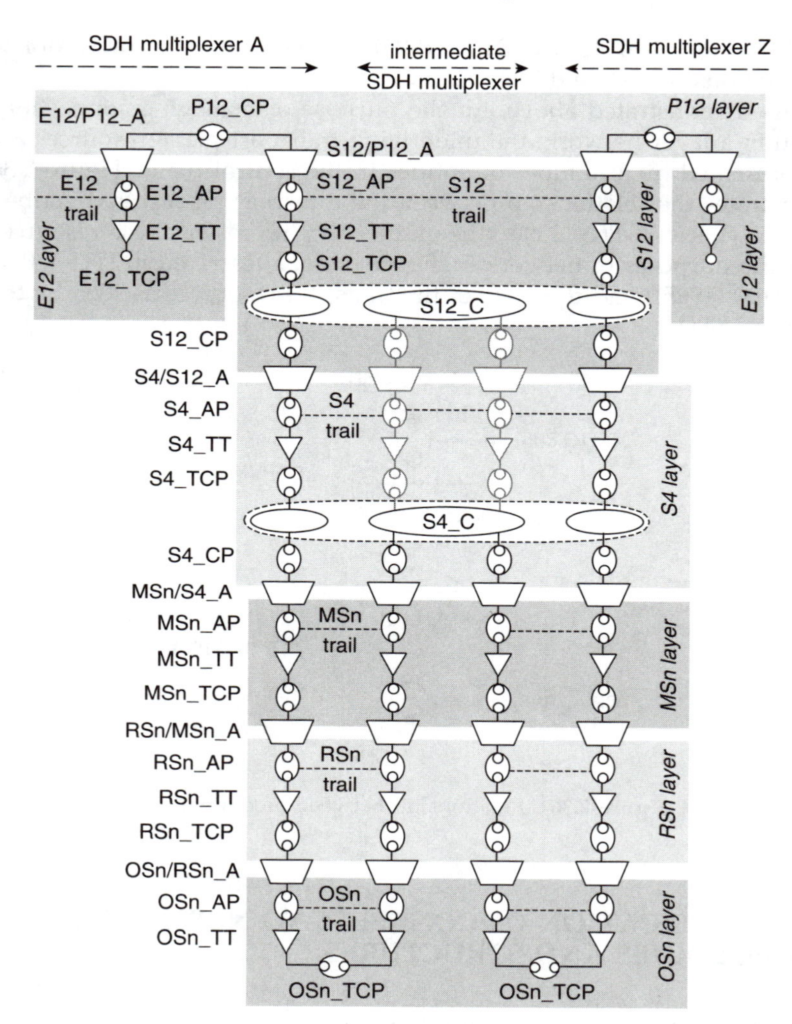

Figure 2.37 PDH transport over SDH network.

(2) The P12 layer network supports the 2048 kbit/s PDH structure as defined in ITU-T Rec. G.702 (1988).
(3) The S12 layer network supports the SDH lower-order path. This layer and the following are defined in ITU-T Rec. G.707 (2003).
(4) The S4 layer network supports the SDH higher-order path.
(5) The MSn layer network supports the SDH multiplex section.

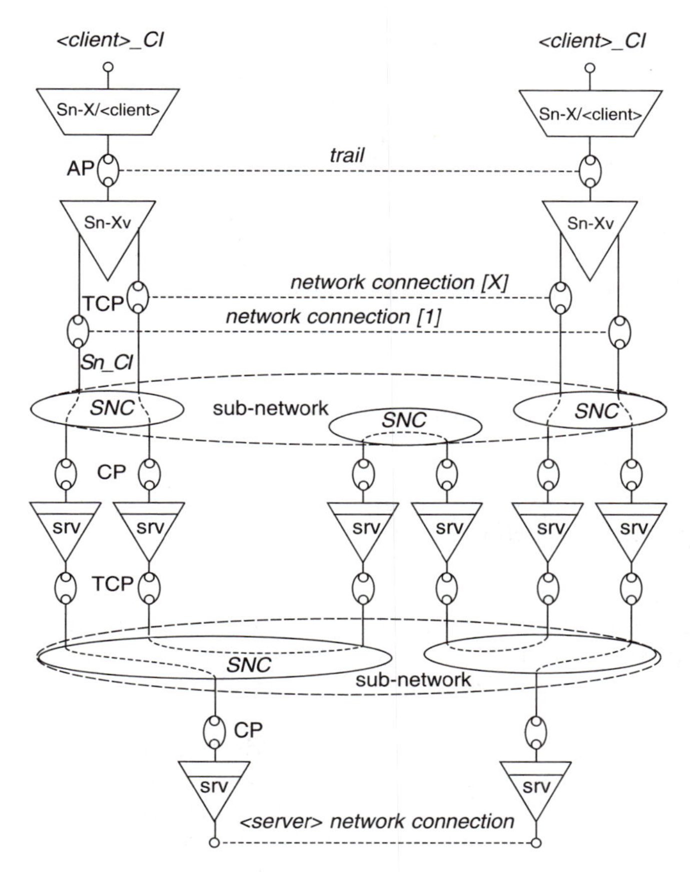

Figure 2.38 Example of inverse multiplexing.

(6) The RSn layer network supports the SDH regenerator section.
(7) The OSn SDH layer network supports the SDH optical section, i.e. the optical signal on the interconnecting fiber.

The example shows the two SDH multiplexers, A and Z, with PDH E12 ports at the tributary side and SDH OSn ports at the SDH line side. The OSn ports are connected by a fiber or, as shown in Figure 2.37, to the OSn ports of one (or more) intermediate SDH multiplexers that either cross-connect the lower-order path signal S12 after terminating the higher order S4 path, or cross-connect the higher-order path signal S4. Figure 2.37 contains all the names of the atomic functions, i.e. the termination functions <layer>_TT, the adaptation function <server_layer>/<client_layer>_A and the connection function <layer>_C.

Figure 2.37 also contains the names of the reference points, i.e. the access points <layer>_AP, the termination connection points <layer>_TCP and the connection points <layer>_CP. The access points are the end points of the trails. The S12 trail runs from node A to node Z, the S4 trail runs between the access points associated with the S4_TT functions, and the MSn trails, RSn trails and OSn trails run between adjacent nodes.

2.8.2 *Inverse multiplexing transport*

Figure 2.38 shows an example of the case where the transport of a client signal is supported through inverse multiplexing over a number (X) of parallel Sn network connections. Four layer networks are shown:

(1) the client layer network;
(2) the Sn layer network providing the inverse multiplexing;
(3) the first server layer, e.g. the SDH MSn/RSn/OSn layer network;
(4) a second server layer, e.g. the OTN layer network;

Figure 2.38 depicts the possibility of diverse routing of inverse multiplexed signals. For example, network connection [1] is supported by both SDH and OTN network connections and network connection [X] is supported by two consecutive SDH network connections that are connected in an intermediate node by an Sn sub-network connection.

3

Partitioning and layering

Cutting the cake

Based on the results of the study of telecommunication networks that existed in 1995 and also on the fact that the new SDH and SONET transport networks have a hierarchical structure, the model that has been selected to represent a telecommunication transport network is a layered model. Partitioning is used to introduce details in a layer.

3.1 LAYERING CONCEPT

In the layered model a transport network can be represented by a number of independent layers. Each layer in this model is used to identify the transport of information of a specific nature. To indicate that a specific layer is used to represent a set of connections within the overall network, it is generally referred to as a *layer network*. The hierarchical structure present in the total network means that each of the layer networks can use the transport capabilities of the underlying layer network. Therefore, a client/server relationship exists between adjacent layer networks in the overall network as illustrated in Figure 3.1. The definition of a particular layer network contains a description of the generation, the transport and the termination of a specific type of information. In the functional model this specific type of information is referred to as characteristic information.

SDH/SONET Explained in Functional Models Huub van Helvoort

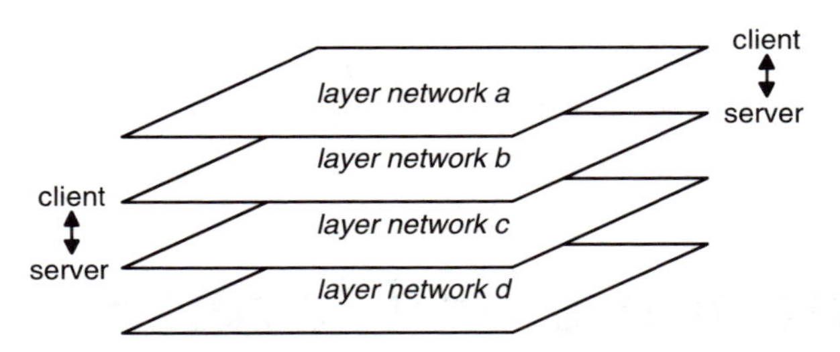

Figure 3.1 Network layers.

The layer networks that have been identified in the transport network functional model should not be confused with the layers that are used in the OSI Model (ITU-T Rec. X.200 (1994)). By definition an OSI layer offers a specific service using one particular protocol out of a set of protocols. This is completely different from the definition of a layer in a transport functional model where each layer network offers the same service, i.e. generation, transportation and termination, by using a specific protocol, i.e. the characteristic information.

Each individual layer network can be specified by a description of its internal architectural structure. The layer network structure can be defined according to the physical properties (e.g. connectors, fibers, cross-connects) or according to manageable properties (e.g. fixed connections, provisionable connections). The division of a single layer network into the different architectural elements is generally referred to as partitioning. Figure 3.2 shows an example of a partitioned connection oriented layer network with the following architectural elements:

- Access groups, i.e. the input/output ports of a layer network.
- Sub-networks, i.e. the provisionable connection functions providing connectivity between input and output connection points.
- Links, i.e. the fixed connections used to interconnect access groups and sub-networks.

Figure 3.2 can be used to show the partitioning of a connectionless layer network with the following architectural components:

- Access groups, i.e. the ingress/egress ports of a layer network.
- Flow domains, i.e. the functions that transfer information between ingress and egress flow points.

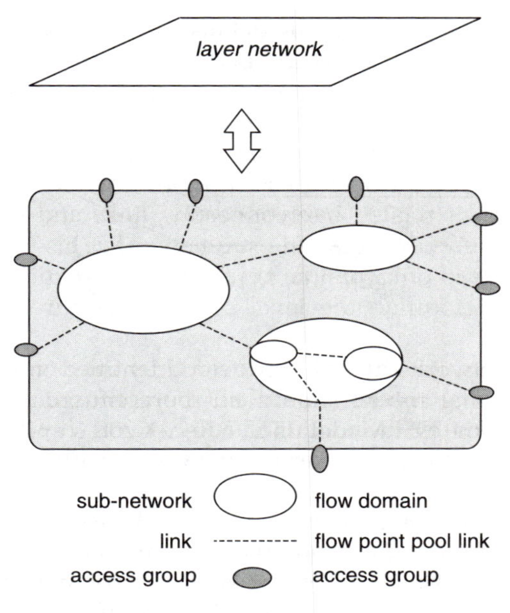

Figure 3.2 Layer partitioning.

- Flow point pool link, i.e. the connectionless means of transferring information between access groups and flow domains.

In the hierarchical transport network the layering concept is a vertical division and the partitioning concept of each layer network is a horizontal division and therefore both concepts are orthogonal.

3.2 PARTITIONING CONCEPT

As described above each layer network contains three architectural structures, i.e. access groups, sub-networks or flow domains and links. Together they define the topology of a layer network or part of it. The partitioning of each structure is described below.

3.2.1 Sub-network partitioning

The smallest sub-network structure is a switch matrix because from a functional point of view a matrix cannot be partitioned any further without showing implementation details. Since a model will represent the functionality of a structure and not its implementation, the sub-network is one of the elements used in the definition of a layer network. In a layer network the links are used to interconnect sub-networks, access groups or sub-networks and access groups. A sub-network may represent two or more contained sub-networks together with their related interconnecting links and is consequently termed a compound or containing sub-network. This representation is recursive. The level of recursion, i.e. the level of partitioning of a sub-network, will determine the level of detail required by a specific representation.

The interconnection capability provided by the ports at the boundary of a containing sub-network shall represent completely the connectivity provided by the contained sub-networks and links.

Thus the general rule for partitioning a sub-network is:

> Any (containing) sub-network may be partitioned into a number of smaller (contained) sub-networks that are interconnected by links. The partitioning of a sub-network shall represent its connectivity exactly, i.e. without increasing or decreasing its interconnection capability.

An example of partitioning sub-networks is the division of a layer network into an international sub-network and several national sub-networks and interconnecting links. Figure 3.3 shows the contained sub-networks of a transport network extended over multiple countries.

Each of these contained (inter-)national sub-networks can be partitioned further to provide more detail. This is illustrated in Figure 3.4 where the contained sub-networks are partitioned into transit sub-networks T and access sub-networks A and interconnecting links, etc.

The layer network modeled in these figures shows all possible input and output ports of the total network and provides a complete overview of the total connectivity of the network. However, for certain representations only a simple model is required. This is the case when only a single connection in the network is regarded and a one-dimensional representation will suffice. The relation between the two-dimensional and one-dimensional representation of the network is depicted in Figure 3.5.

A network connection in a sub-network can be drawn in more detail by showing the contained sub-networks with their own sub-network connections, the interconnecting link connections and the connection

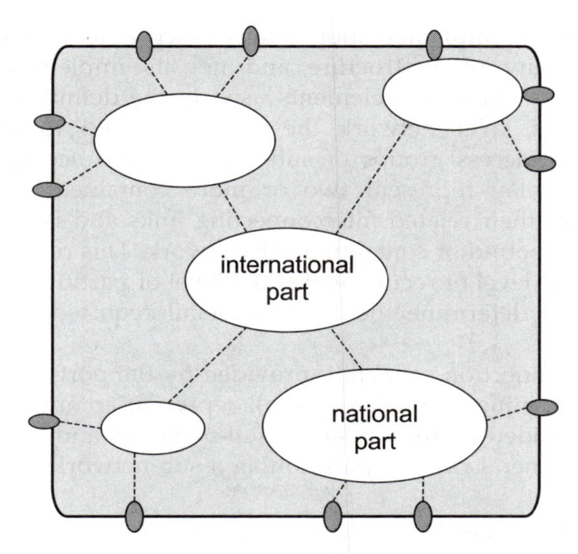

Figure 3.3 Partitioned layer network.

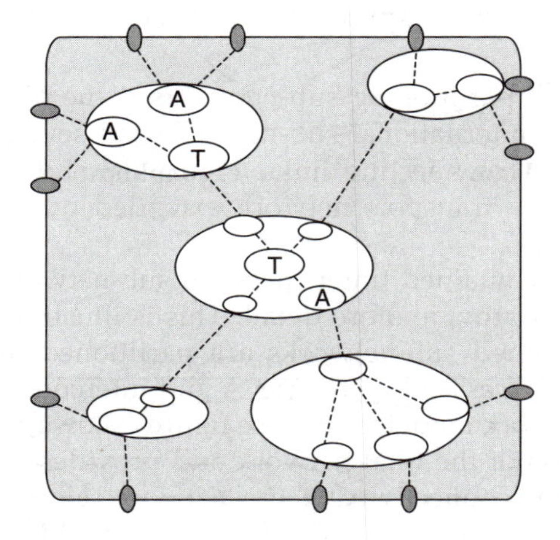

Figure 3.4 Further partitioned layer network.

points of links and sub-networks. The termination connection points identify the terminations or endpoints of a network connection. See Figure 3.6 for a model representation of a network connection.

The recursive nature of the partitioning concept allows the partitioning of a network connection or a sub-network connection into a

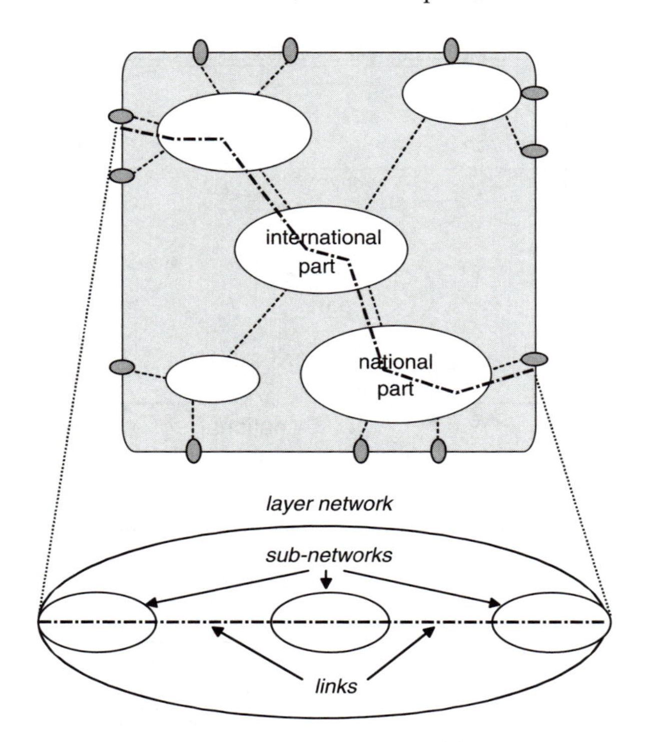

Figure 3.5 Layer model simplification.

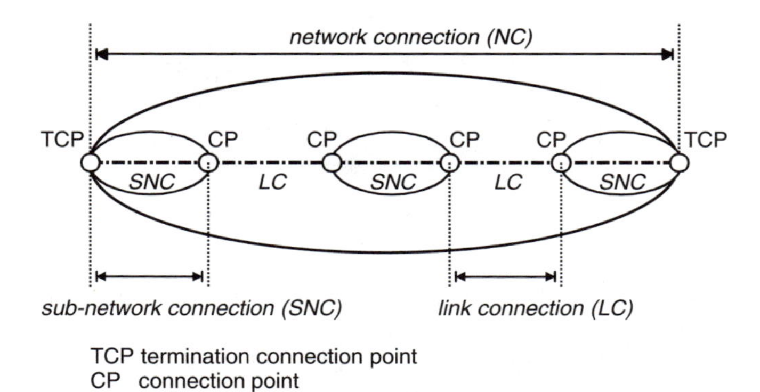

Figure 3.6 Layer network connections.

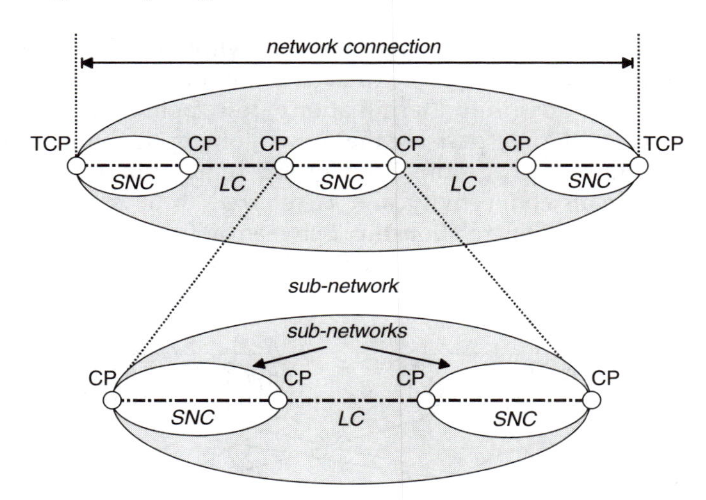

Figure 3.7 Recursive sub-network partitioning.

concatenation of link connections and/or (contained) sub-networks with their own sub-network connections that reflect the partitioning of a sub-network. This is illustrated in Figure 3.7. This partitioning can continue until a single link connection or a matrix remains.

3.2.2 *Flow domain partitioning*

Generally, in a flow domain any ingress flow point can be associated with any egress flow point. Figure 3.8 shows an example

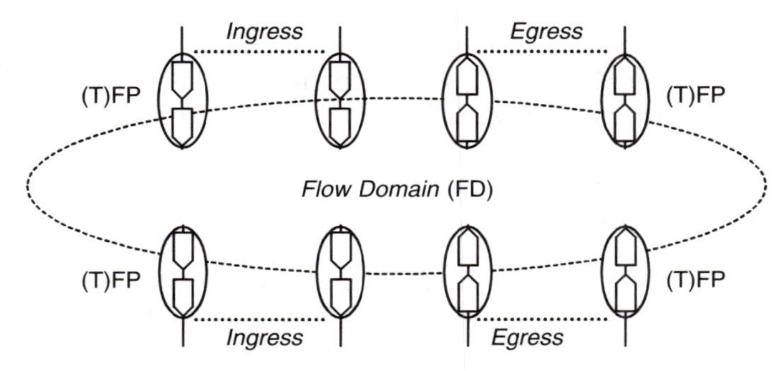

Figure 3.8 Unrestricted flow domain.

representation of a flow domain and its complete set of input and output (termination) flow points that provide full connectivity.

It is possible to group (termination) flow points such that the connectivity within a part of the flow domain is limited to the members within each group. Each group represents a fragment of the flow domain connectivity and is referred to as a flow domain fragment (FDFr). The relationship between a flow domain and its fragments is illustrated in Figure 3.9.

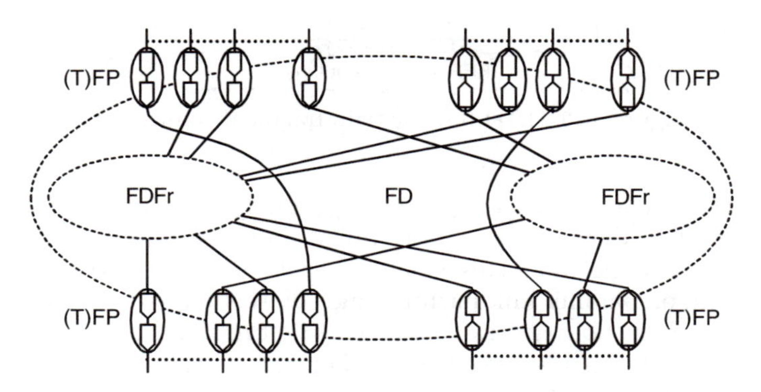

Figure 3.9 Flow domain partitioning.

This partitioning can be applied to any flow domain. When used in a matrix, the fragments are referred to as matrix fragments. A flow domain fragment may be identified by its associated layer network name, e.g. ETH_FDFr, its fragment number, e.g. FDFr_x, or by means of grouping flow points into a characteristic fragment, e.g. in the ETH layer network by the VLAN identifier.

A fragment of one flow domain can be associated with a fragment in another flow domain by means of an interconnecting Flow Point Pool component link or FPP link, as illustrated in Figure 3.10.

3.2.3 *Link partitioning*

Generally, a link in a transport network model represents one or more link connections. A link connection is the smallest unit of manageable capacity that is capable of transporting or routing the characteristic information in a specific layer network. Depending on the required

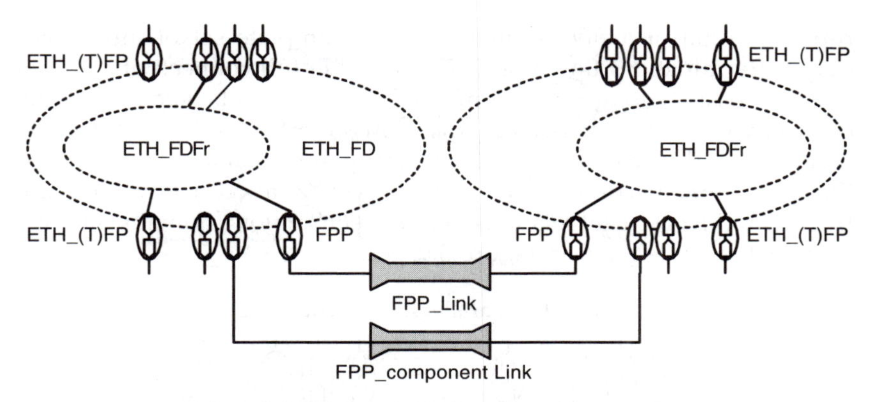

Figure 3.10 ETH flow domain fragment connections.

level of detail shown in a layer network model, and to keep a model simple and readable, two or more links can be bundled together and drawn as a single link. This is shown in Figure 3.11 and sometimes referred to as parallel link partitioning.

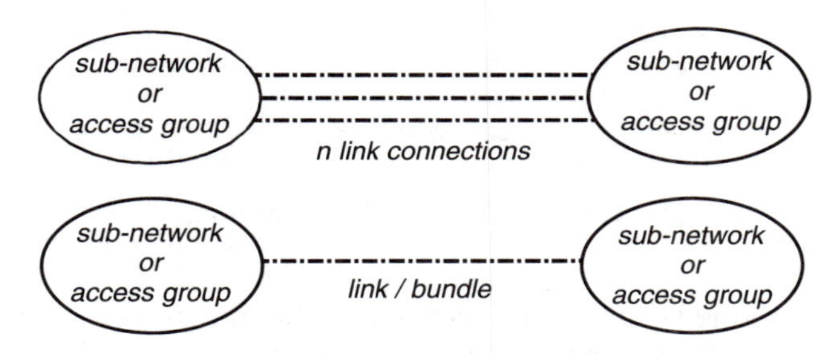

Figure 3.11 Parallel link partitioning.

In connectionless layer networks, parallel partitioning is also used for flows where a set of Flow Points (FP) is grouped in a Flow Point Pool (FPP) and two FPPs are linked by an FPP link (see Figure 3.12). An ETH flow point pool link can be partitioned into ETH flow point pool component links (cLink). There is a maximum to the number of ETH FPP cLinks within an ETH FPP link that can be supported with the VLAN technology.

Apart from links representing parallel links, a link can also be used in a layer network model to represent the concatenation of a link

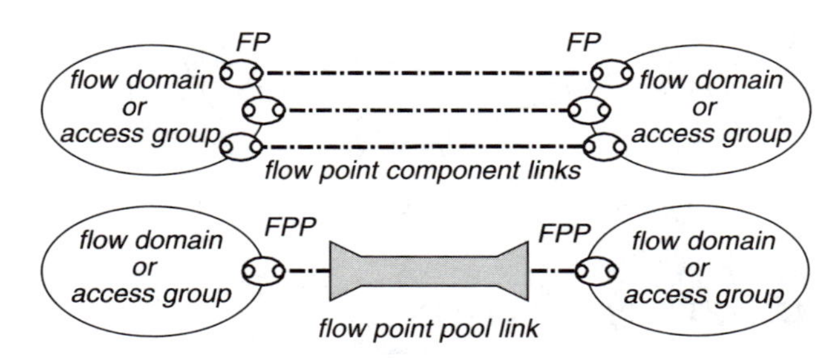

Figure 3.12 Parallel flow partitioning.

connection, a sub-network containing a sub-network connection and a link connection. Such a link is sometimes referred to as a compound link or a serial partitioned link. An example is shown in Figure 3.13.

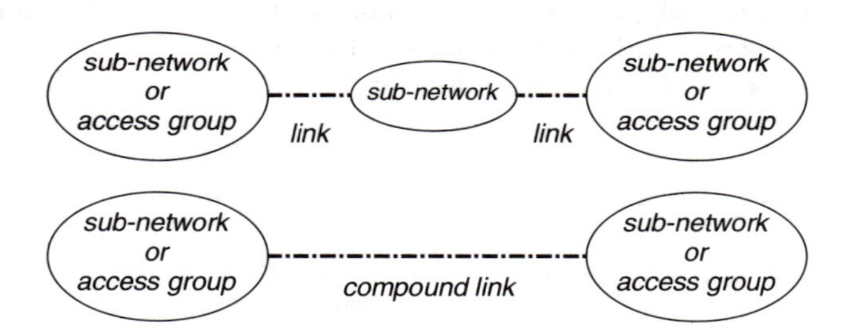

Figure 3.13 Serial link partitioning.

The partitioned links may themselves be recursively partitioned further. This partitioning can continue until a single (physical) link connection remains.

3.2.4 *Access group partitioning*

By definition an access group is a group of co-located trail termination functions together with their associated access points that are connected to the same sub-network or link. Partitioning an access group will result in one or more individual trail termination functions; because these are atomic functions no further partitioning can take place. This is shown in Figure 3.14.

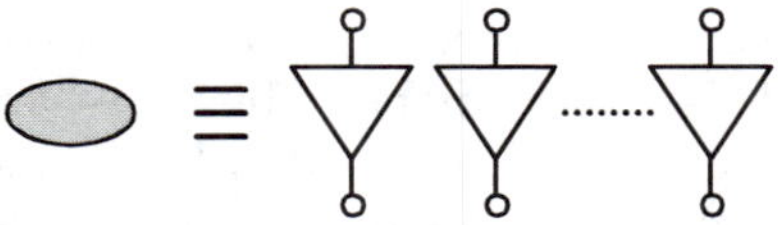

Figure 3.14 Access group partitioning.

3.3 CONCEPT APPLICATIONS

3.3.1 Application of the layering concept

The layering concept applied to the transport network allows us to:

(1) Describe and identify each of the individual layer networks that is part of an overall telecommunication network, by using a similar set of functions.
(2) Define, operate and manage each of the individual layer networks independently of any other layer network.
(3) Assign to each individual layer network its own specific Operations, Administration and Maintenance (OA&M) capabilities, e.g. performance monitoring and automatic failure recovery.
(4) Add, remove or modify each of the individual layer networks without affecting any of the other layer network from an architectural point of view.
(5) Model telecommunication networks that contain multiple transport technologies in a clear and uncomplicated way.

3.3.2 Application of the partitioning concept

The partitioning concept is important as a framework for defining and identifying:

(1) The actual or the functional network structure within each of the layer networks that constitute the overall telecommunications transport network.
(2) The administrative boundaries between sub-networks managed by individual network operators. The complete set of sub-networks will provide all the possible connections within a single layer network. In the ideal case this could be a global layer network.

(3) The boundaries determined by the proportional division and distribution of the performance objectives among the architectural components of a layer network, or sub-network, that is under the responsibility of a single network operator.
(4) The boundaries of the routing domain within a layer network that is owned by a single network operator.
(5) The boundaries of the routing domain within a layer network, or sub-network, that is under the control of a third party (e.g. customer network management).

4

Expansion and reduction

More or less detailed

Similar to the partitioning of a network layer in order to provide more detail by introducing sub-networks, it is also possible to partition, or expand, an atomic function if more detail is required. This expansion could be required to support the explanation or description of certain detailed aspects of a functional model. Expansion can also be used to increase the functionality provided in the network by inserting a new sublayer. In fact expansion is a method to decompose a layered network. In this way fault recovery and monitoring arrangements can be depicted.

Conversely, if the level of detail is such that it distracts from the objective of a provided functional model, two or more atomic functions can be combined and represented by a compound function that will provide the same functionality. Compound functions will also help to reduce the complexity of a model.

4.1 EXPANSION OF LAYER NETWORKS

In a telecommunications network, two specific types of layer networks can be distinguished:

- A path layer network that is independent of the transmission media and dedicated to the transfer of characteristic

SDH/SONET Explained in Functional Models Huub van Helvoort

information between path layer network access points at its boundary.
- A transmission media layer network that may be media dependent and that is dedicated to the transfer of information between transmission media layer network access points at its boundary. A transmission media layer network will support of one or more 'path layer networks'.

The expansion of these layer networks is described in the following section.

4.1.1 Expansion of the path layer network

It is possible to identify in the total path layer network a set of specific path layer networks that are likely to be managed independently by a network operator. Each specific path layer network of the set has its characteristic information transfer capability. This capability can be used either to support the transport of various types of services or support other specific path layer networks as clients. A specific path layer network can have either the transmission media layer network or other specific path layer networks as servers. The actual expansion used to generate the set of specific path layer networks is technology dependent. Each specific path layer network can have a topology independent of other layer networks and it should be an objective that paths across a specific path layer network can be set up independently from the setup of paths in other specific path layer networks. Examples of the decomposition of the path layer network are given in Chapter 11.

4.1.2 Expansion of the transmission media layer

It is possible to identify in the overall transmission media layer network a set of specific transmission media layer networks that are likely to be managed independently by a network operator. The connectivity present in a transmission media layer network cannot be modified directly by management action but most likely will require a physical action. This is based on the current state of technology; in the future it may be possible that the connectivity in a transmission media layer network will become provisionable. Transmission media layer networks can be sub-divided into section layer networks and physical media layer networks.

The section layer networks contain all the functions required for the transfer of characteristic information between physical locations in path layer networks. The section layer network may be expanded into specific section layer networks as described in the examples in Chapter 11.

The physical media layer networks consist of the actual optical fibers, metallic wires or radio frequency channels to serve a section layer network. The physical media layer network may be expanded into specific physical media layer networks to represent, for example, wave division multiplexing. Since the physical media layer network is generally considered to be the lowest layer network, it does not have a server layer network. Hence a physical media layer network connection is supported directly by the media and not by a server trail.

4.1.3 Expansion of specific layer networks into sublayers

Sometimes it is useful to show more functional details in a specific layer network. This can be achieved by introducing sublayers that support specific transport processing functions so that reference points can be shown. A sublayer can be created by expanding the trail termination functions or connection points of a specific layer network.

A sublayer is an integral part of a specific layer network. The difference between a sublayer and a specific layer network is that the functionality provided by a sublayer is neither visible to the client layer network nor to the server layer network of the associated encapsulating specific layer network. A sublayer offers no transport service to a client layer network.

Examples of expansion into sublayers are:

- trail protection schemes by the expansion of the trail termination (see Chapter 9, Section 9.2.1);
- sub-network connection protection schemes by the expansion of the connection point (see Chapter 9, Section 9.2.2);
- tandem connection sublayers to monitor a part of a trail by the expansion of the connection point (see Chapter 8, Section 8.3.2).

The expansion of the trail termination function and connection point is illustrated in the next section.

Sublayering may be used recursively.

4.2 GENERAL PRINCIPLES OF EXPANSION OF LAYERS

It is possible to expand a layer network by expanding either the trail terminations or the (termination) connection points of the layer network. Also, the adaptation functions between two adjacent layer networks can be expanded to create a new layer network.

4.2.1 *Adaptation expansion*

This technique allows specifying the adaptation of the <client> characteristic information to the <server> characteristic information in more detail. It can be depicted by introducing an extra <server′> layer with its particular <server′>_TT and <server′>/<client>_A functions, shown as shaded in Figure 4.1.

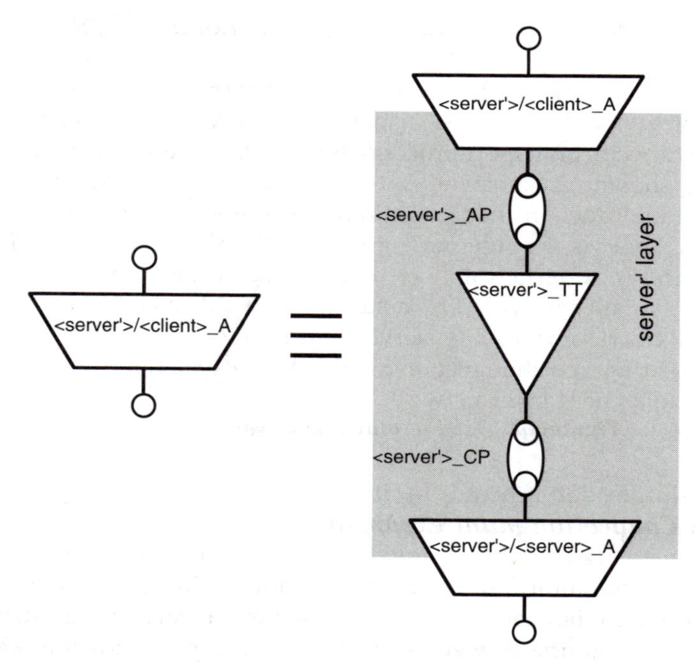

Figure 4.1 Adaptation function expansion.

4.2.2 *Trail termination expansion*

This technique allows specifying the termination of the <client> characteristic information in more detail. It can be depicted by introducing an extra <server′> layer with its particular <server′>/<client>_A and <server′>_TT functions, shown as shaded in Figure 4.2.

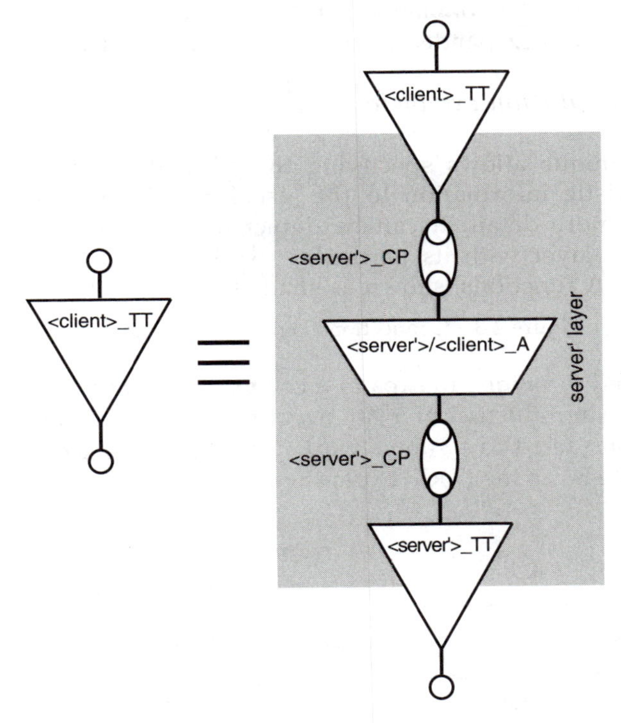

Figure 4.2 Trail termination function expansion.

4.2.3 *Connection point expansion*

This technique allows the introduction of additional resources. It can be depicted by introducing an extra <server> layer with its particular <server>/<client>_A and <server>_TT functions, shown as shaded in Figure 4.3.

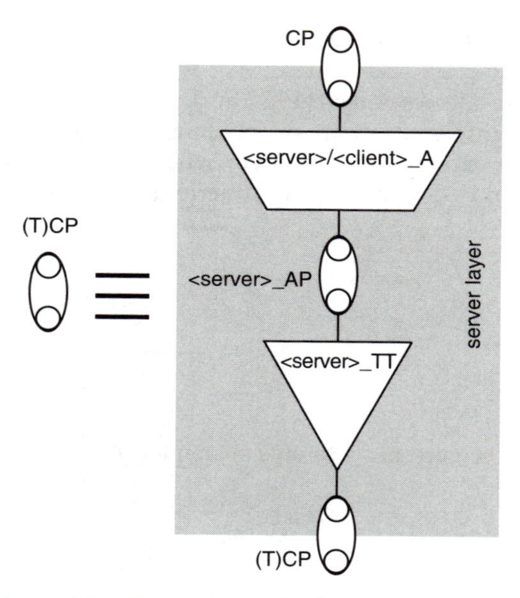

Figure 4.3 Connection point function expansion.

A second technique to expand a connection point is in fact equivalent to the introduction of a sub-network because a connection point can be considered as a sub-network connection with a fixed binding. Figure 4.4 shows the (T)CP expansion.

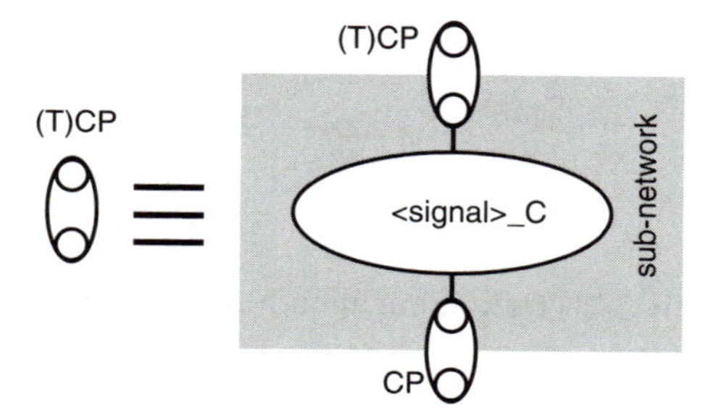

Figure 4.4 Connection point function expansion (second technique).

4.3 REDUCTION OF DETAIL

Figure 4.5 shows the simplest example of the reduction of detail that can be used in a functional model. The resulting symbol and

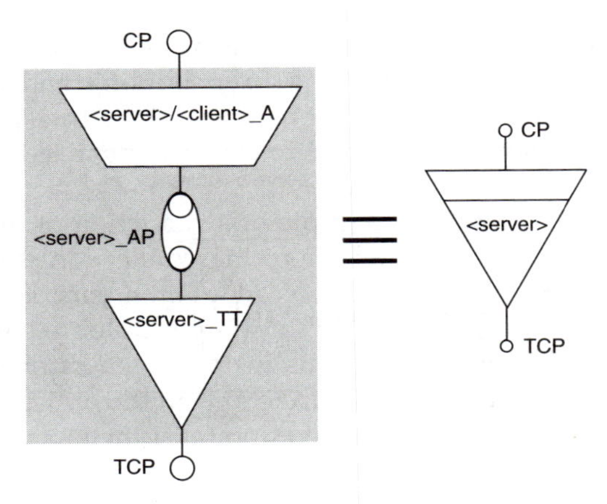

Figure 4.5 Example of simple reduction.

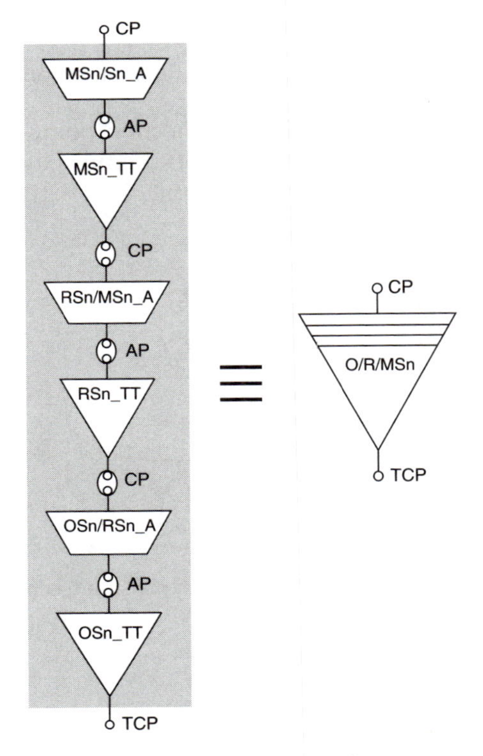

Figure 4.6 Example of reduction.

functionality is commonly referred to as a *compound function*. In this case the access point is normally a fixed binding and the sequence of adaptation function <server>/<client>_A, access point AP and termination function <server>_TT can be depicted by using a <server>_T compound function.

Figure 4.6 shows another example of a reduction of detail that can be used in a functional model. Here the connection points are also fixed bindings and the sequence of adaptation function MSn/Sn_A and termination function MSn_TT in the multiplex section layer, the Rsn/MSn_A and RSn_TT functions in the regenerator section layer and the OSn/RSn_A and Osn_TT functions in the optical section layer can be replaced by an O/R/MSn compound function.

5

Adaptation functions

The glue between layer networks

This chapter describes the generic atomic model that represents the adaptation of client layer network signals for transport over server layer networks. Hence it is termed an adaptation function. Although the generic model is fairly simple the level of detail in the atomic function can be raised by applying the partitioning methodology described in Chapter 4, Section 4.2.1. Examples of existing adaptation models are provided to exercise the application of adaptation functions.

5.1 GENERIC ADAPTATION FUNCTION

The generic adaptation function is the atomic function that performs the adaptation of a signal in a client layer network to a format that is suitable for the transport over a trail in a server layer network. Consequently, at the far end of the trail, this fnction will recover the original client signal for further transport in the client layer. In the following figures, <client> represents the notation used for the client signal, e.g. MSn, Sn, OTUk, ODUk, etc., and <server> represents the server signal, e.g. RSn, Sn, OCh, etc. The standard convention to identify an adaptation function is: <server>/<client>_A. Figure 5.1 shows the bi-directional functional model.

In the following description of the trail termination function, the same documentation template is used as the template used in most of the ETSI standards and ITU-T Recommendations. It contains separate paragraphs for:

- **symbol**, a figure representing the (atomic) function in a functional model;

SDH/SONET Explained in Functional Models Huub van Helvoort

- **interfaces**, a table containing all input and output signals;
- **processes**, detailed description of the function specific processes;
- **defects**, list of the defects detected by this function, e.g. dLOP;
- **consequent actions**, list of the actions based on the detected defects, e.g. aAIS;
- **defect correlation**, list of the correlations used to determine the most probable cause of defects, e.g. cPLM;
- **performance monitoring**, list of the performance monitoring primitives, e.g. pN_DS.

Bi-directional model: <server>/<client>_A

The connection point <client>_CP represents the binding with the client layer connection function <client>_C described in Chapter 7, Section 7.1. The access point <server>_AP represents the binding with the server layer trail termination function <server>_TT described in Chapter 6, Section 6.1. The adaptation management point <server>/<client>_A_MP represents the binding with the management system and is used to provision the adaptation function and receive information from the function. This model can be decomposed into the uni-directional models depicted in Figures 5.2 and 5.4.

Symbol

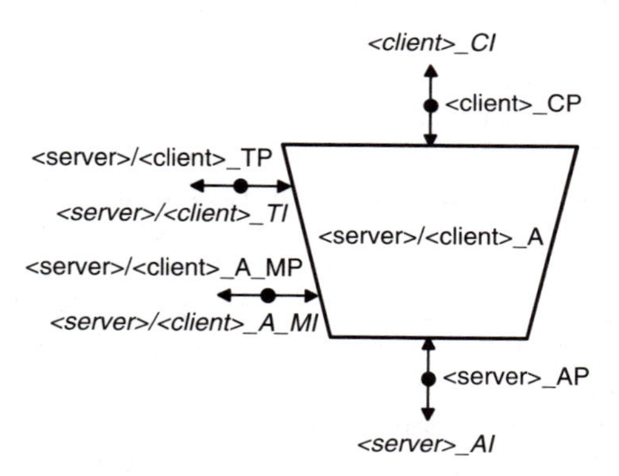

Figure 5.1 Bi-directional <server>/<client>_A function.

Source side: <server>/<client>_A_So

This is the uni-directional function that performs the adaptation of the client signal with its client layer network characteristic information <client>_CI to the server layer network adapted information signals

<server>_AI transported over the network layer trail between the server layer network access points <server>_AP. This function is located at the beginning or source side of the network trail.

The information flow and the processing of the <server>/<client>_A_So function are defined with reference to Figures 5.2 and 5.3.

Symbol

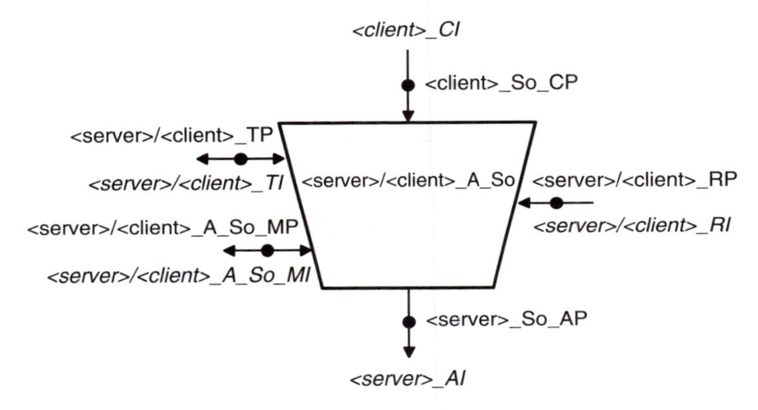

Figure 5.2 Source side <server>/<client>_A function.

Interfaces

<server>/<client>_A_So input and output signals.

inputs	outputs
at <client>_So_CP client signal characteristic information <client>_CI_nn **at <server>/<client>_A_So_MP** server/client adaptation management information <server>/<client>_A_So_MI_nn **at <server>/<client>_TP** server/client adaptation timing information <server>/<client>_TI_nn	**at <server>_So_AP** server signal adapted information <server>_AI_nn **at <server>/<client>_A_So_MP** server/client adaptation management information <server>/<client>_A_So_MI_nn **at <server>/<client>_RP** server/client adaptation remote information <server>/<client>_RI_nn

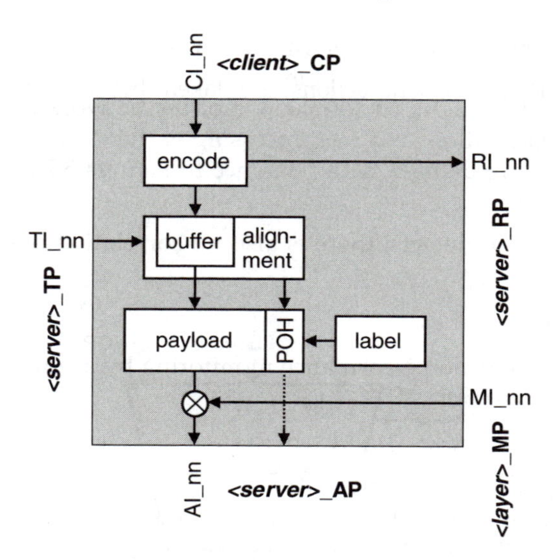

Figure 5.3 <server>/<client>_A_S processes.

Processes

- ***Activation***: The <server>/<client>_A_So function shall access the access point when it is activated (MI_Active is true). Otherwise, it shall not access the access point and the following processes will not be activated.
- ***Client data stream encoding***: The client characteristic information is adapted to the server characteristic structure.
- ***Client alignment***: The adaptation source aligns the client data stream with the server layer structure and adds alignment information. While the alignment process is technology dependent, time division multiplexed (TDM) systems commonly require buffering of the signal. Synchronization information is retrieved from the timing reference point.
- ***Client labeling***: Each client signal is 'labeled' to enable multiplexing of client data streams and to identify the type of multiplexing because the multiplexing methodology is very technology specific.

Defects

In general, no defects are detected by a source adaptation function.

Consequent actions

In general, no consequent actions are taken by a source adaptation function.

Defect correlation

Since there are no defects, there is also no correlation of defects.

Performance monitoring

In general, there is no performance monitoring by a source adaptation function.

Sink side: <server>/<client>_A_Sk

This is the uni-directional function that performs the adaptation of the client signal with its client layer network characteristic information <client>_CI to the server layer network adapted information signals <server>_AI transported over the network layer trail between the server layer network access points <server>_AP. This function is located at the end or sink side of the network trail. The information flow and the processing of the <server>/<client>_A_Sk function are defined with reference to Figures 5.4 and 5.5.

Symbol

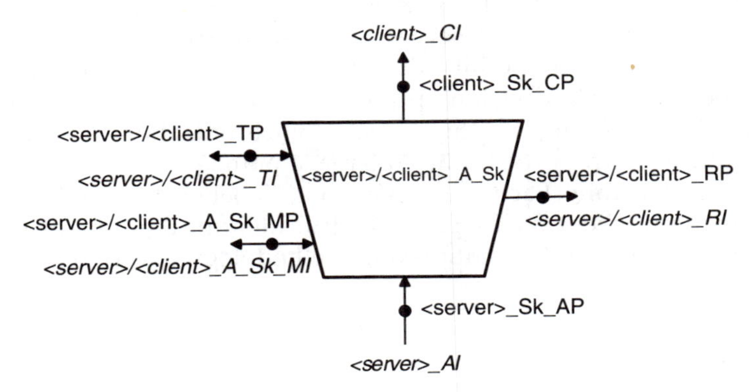

Figure 5.4 Sink side <server>/<client>_A function.

Interfaces

<server>/<client>_A_Sk input and output signals.

inputs	outputs
at <server>_Sk_AP server signal adapted information <server>_AI_nn	**at <client>_Sk_CP** client signal characteristic information <client>_CI_nn
at <server>/<client> A_Sk_MP server/client signal management information <server>/<client>_A_Sk_MI_nn	**at <server>/<client>A_Sk_MP** server/client signal management information <server>/<client>_A_Sk_MI_nn
at <server>/<client>_RP server/client adaptation remote information <server>/<client>_RI_nn	
	at <server>/<client>_TP server/client adaptation timing information <server>/<client>_TI_nn

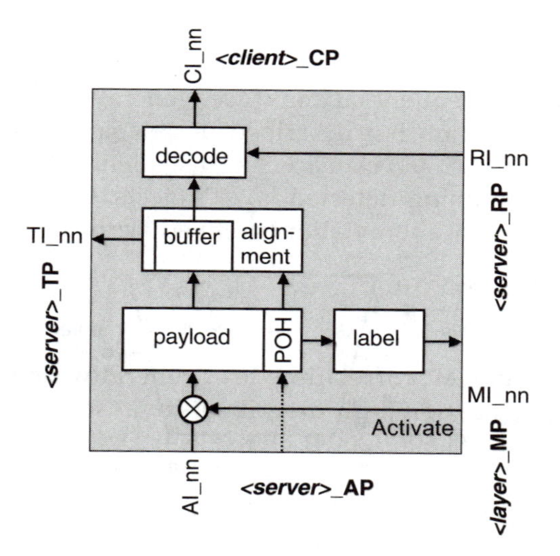

Figure 5.5 <server>/<client>_A_Sk processes.

Processes

- ***Activation***: The <server>/<client>_A_Sk function shall access the access point and perform the common and specific processes operations specified below when it is activated (MI_Active is true). Otherwise, it shall activate the SSF signals at its output (CP) and not report its status via the management point.
- ***Client labeling***: The recovered signal 'label' is used to identify correctly each client signal and the methodology that has to be used to demultiplex it. This demultiplexing is very technology specific.
- ***Client alignment***: The effect of the source alignment is removed by using the alignment information. While the alignment process is technology dependent, time division multiplexed (TDM) systems commonly require buffering of the signal. The recovered client clock is output to the timing reference point.
- ***Client information decoding***: The original client data stream is recovered.

Defects

Definitions of defect processing are generally not part of the functional specification but described in a separate document, e.g. in ITU-T Rec. G.806 (2004) clause 6.2. Defects are identified by an abbreviation prefixed with a 'd': dXXX, e.g. dLOP.

Consequent actions

Definitions of consequent actions are generally not part of the functional specification but described in a separate document, e.g. in ITU-T Rec. G.806 (2004) clause 6.3. Consequent actions are defined by equations containing detected layer defects. Consequent actions are identified by an abbreviation prefixed with an 'a': aXXX, e.g. aSSF.

Defect correlation

Definitions of defect correlation are generally not part of the functional specification but described in a separate document, e.g. in ITU-T Rec. G.806 (2004) clause 6.4. Defect correlations are defined by equations containing detected layer defects; server layer defects and reporting can be disabled by provisioning. Defect correlations are used to find a probable cause and are identified by an abbreviation prefixed with a 'c': cXXX, e.g. cPLM.

Performance monitoring

In general, performance processing is not performed in an adaptation function. Performance primitives are identified by an abbreviation prefixed with a 'p': pXXX, e.g. pFCSError.

When the defects are persistent an alarm is raised and sent to the network management system via the <server>/<client>_A_Sk_MP Management Point.

5.2 ADAPTATION FUNCTION EXAMPLES

There are already many different adaptation functions defined, see ETSI standards EN300 417-1 to 417-7 and 417-9 to 417-10 and ITU-T Rec. G.705 (2000) for PDH, G.783 (2004) for SDH, G.798 (2004) for OTN, I.732 (2000) for ATM, G.8021 (2004) for Ethernet and G.mplseq (2005) for MPLS. In this subsection only a very small subset is described to serve as an example for new developments.

5.2.1 *The Sn/Sm_A function*

The first example is derived from ITU-T Rec. G.783 (2004) clause 12.3.1: the adaptation function Sn/Sm_A that adapts the SDH Sm layer characteristic information for transport in the Sn layer.

(12.3.1) VC–n layer to VC–m layer adaptation Sn/Sm_A

The Sn/Sm_A (m = 11, 12, 2 or 3; n = 3 or 4) provides the primary adaptation functionality between higher order VC–n and lower order VC–m layer networks in SDH. It defines the TU pointer processing, and may be divided into three functions:

- pointer generation at the source side;
- pointer interpretation at the sink side;
- frequency justification at source and sink side.

The S4/S11*_A provides the SDH-SONET interworking functionality for the transport of lower order VC–11s (i.e. VT1.5 SPE) by a VC–4 via TU-12 multiplexing. The * (asterix) indicates that this is a special function because generically a VC–11 is multiplexed in a TU-11. This special function also defines the TU pointer processing, and may be divided into four functions:

- adding at the source and removal at the sink of stuffing bytes;
- pointer generation at the source side;

- pointer interpretation at the sink side;
- frequency justification at the source and sink side.

The format for TU pointers, their roles for processing, and mappings of VCs are described in ITU-T Rec. G.707 (2003).

The Sn/Sm_A function also acts as a source and sink for the overhead octets H4 and C2. The symbol of the bi-directional adaptation model is shown in Figure 5.6.

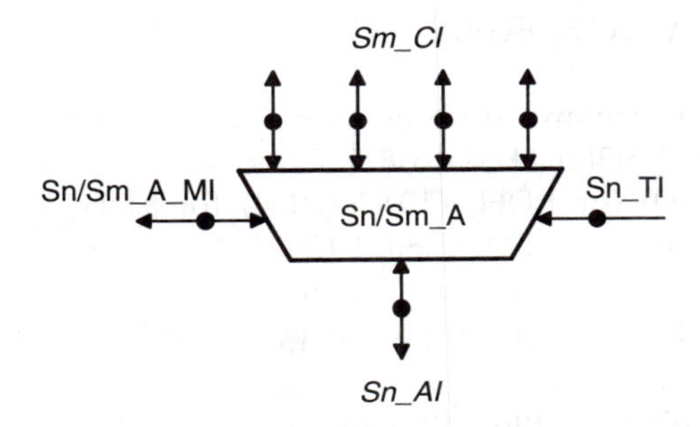

Figure 5.6 Bi-directional Sn/Sm_A function.

This model can be further decomposed into the uni-directional source and sink models. These models are drawn in Figures 5.7 and 5.9.

(12.3.1.1) VC–n layer to VC–m layer adaptation source Sn/Sm_A_So

The information flow and the processing of the Sn/Sm_A_So function are defined with reference to Figures 5.7 and 5.8.

Symbol

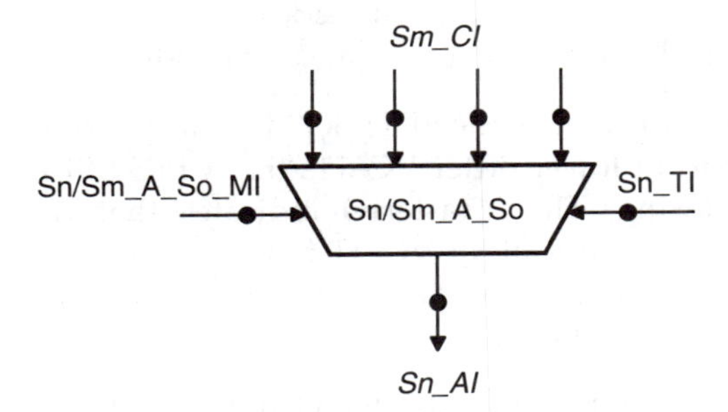

Figure 5.7 Sn/Sm_A_So function.

Interfaces

Sn/Sm_A_So input and output signals.

inputs	outputs
at the Sm_So_CP Sm_CI_Data Sm_CI_Clock Sm_CI_FrameStart Sm_CI_MultiFrameSync **at the Sn_So_TP** Sn_TI_Clock Sn_TI_FrameStart **at the Sn/Sm_A_So_MP** Sn/Sm_A_So_MI_Active	**at the Sn_So_AP** Sn_AI_Data Sn_AI_Clock Sn_AI_FrameStart

Processes

- ***Activation***: The Sn/Sm_A_So function shall access the access point when it is activated (MI_Active is true). Otherwise, it shall not access the access point and the following processes will not be activated.
- ***Assembly***: The Sn/Sm_A function assembles VCs of lower order m (m = 11, 12, 2, 3) as TU-m into VCs of higher order n (n = 3 or 4). In the case of the S4/S11*_A_So function, 36 bytes of fixed stuff are added to the VC–11.

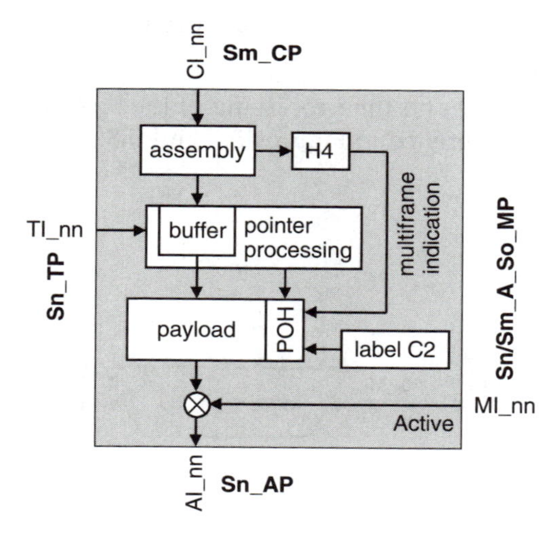

Figure 5.8 Sn/Sm_A_So processes.

- ***Pointer processing***: Consists of:

  - Pointer generation: the frame offset in bytes between a lower order VC and higher order VC is indicated by a TU pointer that is assigned to that particular lower order VC.
  - Frequency justification: Lower order VC data at the Sm_CP are synchronized to the timing information at the Sm_TP reference point. The pointer processing function provides accommodation for wander and plesiochronous offset in the received signal with respect to the synchronous equipment timing reference. This function can be modeled as a data buffer that is being written with data, timed from the received VC clock, and read by a VC clock derived from reference point Sn_TP.

- *H4*: A multiframe indicator is generated and placed in the H4 byte position.
- *C2*: Signal label information derived directly from the adaptation function type is placed in the C2 byte position.

The assembly and pointer processing and the use of H4 and C2 are described in detail in ITU-T Rec. G.707 (2003).

Defects

None.

Consequent actions

The function shall perform the following consequent actions:

$$aAIS \leftarrow CI\_SSF$$

When an all-ONEs (i.e. AIS) signal is detected at the input Sm_CP, an all-ONEs (TU-AIS) signal shall be applied at the output Sn_AP within 2 (multi)frames. Upon clearing of the all-ONEs signal at the Sm_CP, the all-ONEs (TU-AIS) signal shall be removed within 2 (multi)frames.

Defect correlations

None.

Performance monitoring

None.

(12.3.1.2) VC–n layer to VC–m layer adaptation sink Sn/Sm_A_Sk

The information flow and the processing of the Sn/Sm_A_Sk function are defined with reference to Figures 5.9 and 5.10.

Symbol

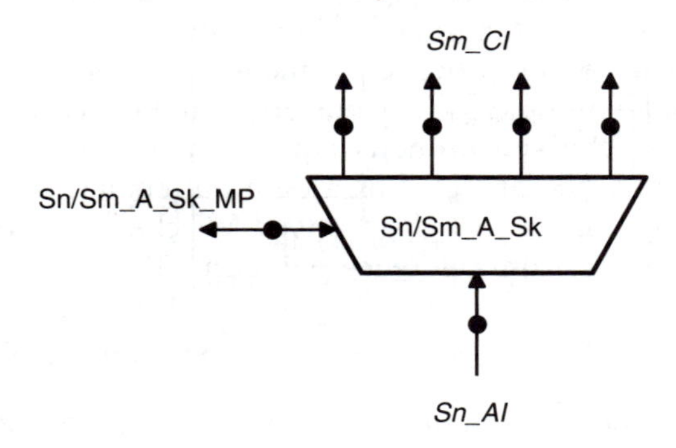

Figure 5.9 Sn/Sm_A_Sk function.

Interfaces

Sn/Sm_A_Sk input and output signals.

inputs	outputs
at Sn_Sk_AP Sn_AI_Data Sn_AI_Clock Sn_AI_FrameStart Sn_AI_TSF	**at Sn_Sk_CP** Sm_CI_Data Sm_CI_Clock Sm_CI_FrameStart Sm_CI_MFS Sm_CI_SSF
at Sn/Sm_A_Sk_MP Sn/Sm_A_Sk_MI_Active Sn/Sm_A_Sk_MI_AIS_Reported	**at Sn/Sm_A_Sk_MP** Sn/Sm_A_Sk_MI_AcSL Sn/Sm_A_Sk_MI_cPLM Sn/Sm_A_Sk_MI_cLOM Sn/Sm_A_Sk_MI_cLOP Sn/Sm_A_Sk_MI_cAIS

Processes

- ***Activation***: The Sn/Sm_A_Sk function shall access the access point and perform the common and specific processes operation

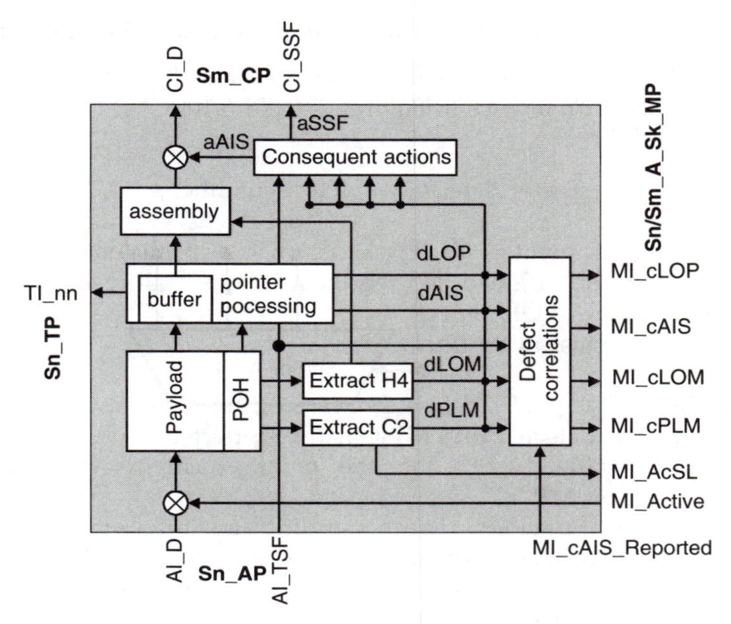

Figure 5.10 Sn/Sm_A_Sk processes.

specified below when it is activated (MI_Active is true). Otherwise, it shall activate the SSF signals at its output (CP) and not report its status via the management point.

- ***Disassembly***: The Sn/Sm_A_Sk function disassembles VC–n (n = 1, 3) into lower order VC–m (m = 11, 12, 2, 3), performing multiframe alignment if necessary. The S4/S11*_A_Sk function strips off the 36 fixed stuff bytes from the VC–12 container to recover the VC–11. Note that this action may cause a discrepancy between performance reports at an S12m_TT_Sk and an S11_TT_Sk for a VC–11 trail.
- ***Pointer processing***: Consists of:

    - Pointer interpretation: The TU pointer of each lower order VC–m is decoded to provide information about the frame offset in octets between the higher order VC and the individual lower order VCs.
    - Frequency justification: This process must allow for continuous pointer adjustments when the clock frequency of the node where the TU was assembled is different from the local

clock reference. The frequency difference between these clocks affects the required size of the data buffer. The TU pointer interpretation is specified in ITU-T Rec. G.783 (2004) Annex A.

The pointer interpreter detects two defect conditions:

- a loss of pointer (dLOP): note that a persistent mismatch between provisioned and received TU type will result in a loss of pointer (dLOP) defect.
- an AIS signal in the TU (dAIS).

- ***C2***: Byte C2 (signal label) is recovered from VC–n overhead. If a mismatch is detected (dPLM) it shall be reported together with the accepted value (AcSL) via the management reference point Sn/Sm_A_Sk_MP.
- ***H4***: When a payload requires multiframe alignment, the H4 byte contains the multiframe indicator and multiframe alignment is performed as defined in ITU-T Rec. G.783 (2004) clause 8.2.2. In this case a loss of multiframe defect (dLOM) shall be detected.

The disassembly and pointer processing and the use of H4 and C2 are described in detail in ITU-T Rec. G.707 (2003).

Defects

The defects dAIS and dLOP shall be detected in this function according to the specification in ITU-T Rec. 783 (2004) Annex A; dLOM and dPLM shall be detected in this function according to the specification in ITU-T Rec. 806 (2004) clause 6.2.

Consequent actions

The function shall perform the following consequent actions as described in ITU-T Rec. G.806 (2004) clause 6.3:
for VC–3

$$\text{aAIS} \leftarrow \text{dPLM or dAIS or dLOP}$$

$$\text{aSSF} \leftarrow \text{dPLM or dAIS or dLOP}$$

for VC–11/VC–12/VC–2

$$aAIS \leftarrow dPLM \text{ or } dLOM \text{ or } dAIS \text{ or } dLOP$$
$$aSSF \leftarrow dPLM \text{ or } dLOM \text{ or } dAIS \text{ or } dLOP$$

Upon the declaration of aAIS, a logical all-ONEs (AIS) signal shall be applied at the Sm_CP within 2 (multi)frames. Upon termination of these aAIS, the all-ONEs signal shall be removed within 2 (multi)-frames.

Defect correlations

The function shall perform the following defect correlations to determine the most probable fault cause as defined in ITU-T Rec. G.806 (2004) clause 6.4. These fault causes shall be reported to the element manger via the Sn/Sm_A_Sk_MP.

$$cPLM \leftarrow dPLM \text{ and } (\text{not } AI\_TSF)$$

for VC–3

$$cAIS \leftarrow dAIS \text{ and } (\text{not } AI\_TSF) \text{ and } (\text{not } dPLM) \text{ and } AIS\_Reported$$
$$cLOP \leftarrow dLOP \text{ and } (\text{not } dPLM)$$

for VC–11/VC–12/VC–2

$$cLOM \leftarrow dLOM \text{ and } (\text{not } AI\_TSF) \text{ and } (\text{not } dPLM)$$
$$cAIS \leftarrow dAIS \text{ and } (\text{not } AI\_TSF) \text{ and } (\text{not } dPLM) \text{ and } (\text{not } dLOM) \text{ and } AIS\_Reported$$
$$cLOP \leftarrow dLOP \text{ and } (\text{not } dPLM) \text{ and } (\text{not } dLOM)$$

Performance monitoring

None.

5.2.2 *The OCh/RSn_A function*

The second example is derived from ITU-T Rec. G.798 (2004) clause 12.3.5: the adaptation function OCh/RSn_A that adapts the SDH

RSn layer characteristic information for transport in the OTN OCh layer. The bi-directional representation of the function is shown in Figure 5.11.

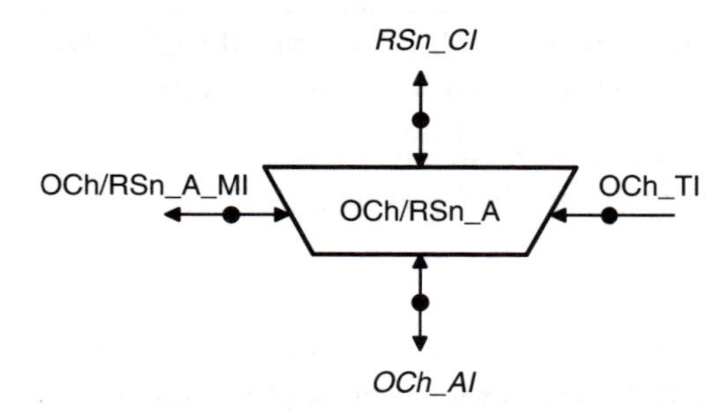

Figure 5.11 Bi-directional Och/RSn_A function.

(12.3.5) OCh to RSn adaptation (OCh/RSn_A)

The OCh to RSn adaptation functions perform the adaptation between the OCh layer adapted information and the characteristic information of a RSn layer signal.

(12.3.5.1) OCh to RSn adaptation source function (OCh/RSn_A_So)

The information flow and processing of the OCh/RSn_A_So function are defined with reference to Figures 5.12 and 5.13.

Symbol

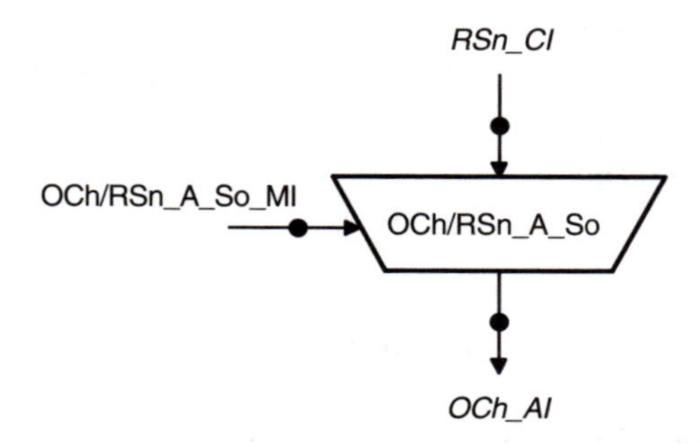

Figure 5.12 Och/RSn_A_So function.

Interfaces

Sn/Sm_A_So input and output signals.

inputs	outputs
at the RSn_So_CP RSn_CI_Data RSn_CI_Clock **at the OCh/RSn_A_So_MP** OCh/RSn_A_So_MI_Active	**at the OCh_So_AP** OCh_AI_Data

Processes

The function generates the OCh_AI signal from the RSn_CI.

- ***Activation***: The OCh/RSn_A_So function shall access the access point when it is activated (MI_Active is true). Otherwise, it shall not access the access point. The jitter and wander requirements are defined in ITU-T Rec. G.783 (2004) clause 9.3.1.1.

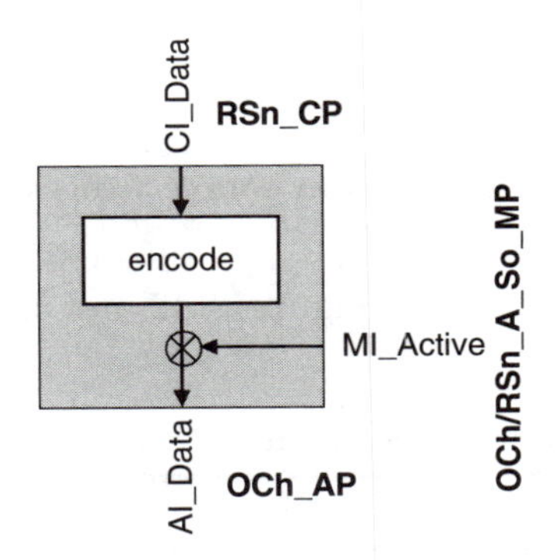

Figure 5.13 Och/RSn_A_So processes.

Defects

None.

Consequent actions

None.

Defect correlations

None.

Performance monitoring

None.

(12.3.5.2) OCh to RSn adaptation sink function (OCh/RSn_A_Sk)

The information flow and processing of the OCh/RSn_A_Sk function are defined with reference to Figures 5.14 and 5.15.

Symbol

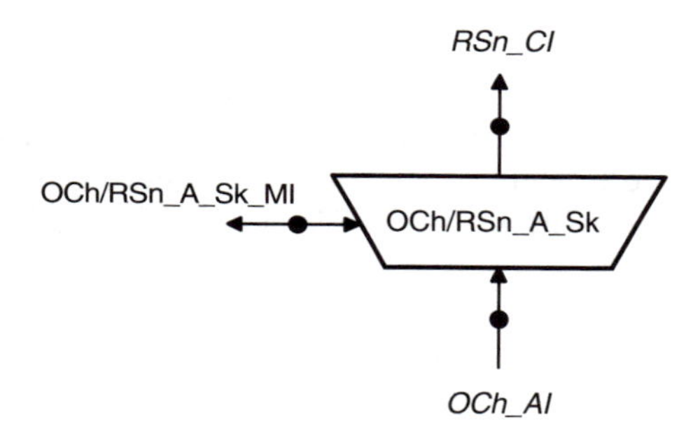

Figure 5.14 Och/RSn_A_Sk function.

Interfaces

OCh/RSn_A_Sk input and output signals.

inputs	outputs
at the OCh_So_CP OCh_CI_Data OCh_CI_TSF **at the Sn_So_TP** Sn_TI_Clock Sn_TI_FrameStart **at the OCh/RSn_A_So_MP** OCh/RSn_A_Sk_MI_Active	**at the RSn_So_AP** RSn_AI_Data RSn_AI_Clock RSn_AI_FrameStart RSn_AI_SSF **at the OCh/RSn_A_Sk_MP** OCh/RSn_A_Sk_MI_cLOF

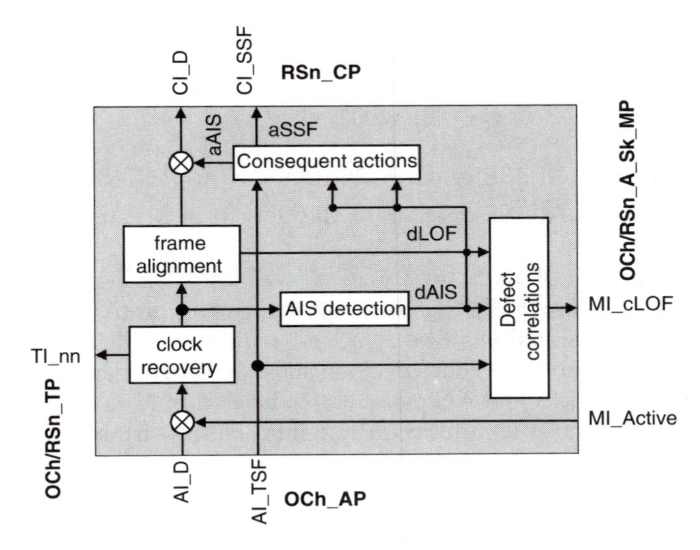

Figure 5.15 Och/RSn_A_Sk processes.

Processes

The processes associated with the OCh/RSn_A_Sk function are depicted in Figure 5.15.

- ***Activation***: The OCh/RSn_A_Sk function shall access the access point and perform the common and specific processes operation specified below when it is activated (MI_Active is true). Otherwise, it shall activate the SSF signals and generate AIS at its output (CP) and not report its status via the management point.
- ***Clock recovery***: The function shall recover the RSn clock signal from the incoming data. The supported input clock range, the immunity against the presence of consecutive identical digits, the intrinsic jitter at the STM-N output and the jitter transfer shall be as specified in ITU-T Rec. /G.783 (2004) clause 15.1.3.
- ***Frame alignment***: The function shall perform frame alignment on the STM-N frame as described in ITU-T Rec. G.783 (2004) clause 8.2.1.

Defects

The defects dAIS and dLOF shall be detected in this function according to the specification in ITU-T Rec. 806 (2004) clause 6.2.

Consequent actions

The function shall perform the following consequent actions as described in ITU-T Rec. G.806 (2004) clause 6.3:

aSSF ← AI_TSF or dAIS or dLOF or (not MI_Active)

aAIS ← AI_TSF or dAIS or dLOF or (not MI_Active)

On declaration of aAIS, the function shall output a logical all-ONEs (AIS) signal within 2 STM-N frames. On clearing of aAIS, the logical all-ONEs (AIS) signal shall be removed within 2 STM-N frames and normal data be output. The AIS clock start shall be independent from the incoming clock. The AIS clock has to be within N × 155 520 kbit/s ± 20 ppm. Jitter and wander requirements are for further study.

Defect correlations

The function shall perform the following defect correlations to determine the most probable fault cause as defined in ITU-T Rec. G.806 (2004) clause 6.4. This fault cause shall be reported to the element manger via the OCh/RSn_A_Sk_MP.

cLOF ← dLOF and (not dAIS) and (not AI_TSF)

(Note: dAIS is not reported as fault cause as it is a secondary alarm and will result in aSSF, which is reported as cSSF fault cause in the RSn_TT_Sk that directly follows this function.)

Performance monitoring

None.

5.2.3 *The LCAS capable Sn–X–L/ETH_A function*

The third example is derived from ITU-T Rec. G.8021 (2004) clause 11.1.2: the adaptation function Sn–X–L/ETH_A that adapts the ETH layer characteristic information for transport in an Sn–X container. This container will be transported in the VC–n layer by X individual VC–n by using LCAS capable virtual concatenation. This is a technology specific application of the model described in Chapter 10, Section 10.2.5. Figure 5.16 shows the bi-directional model of this function.

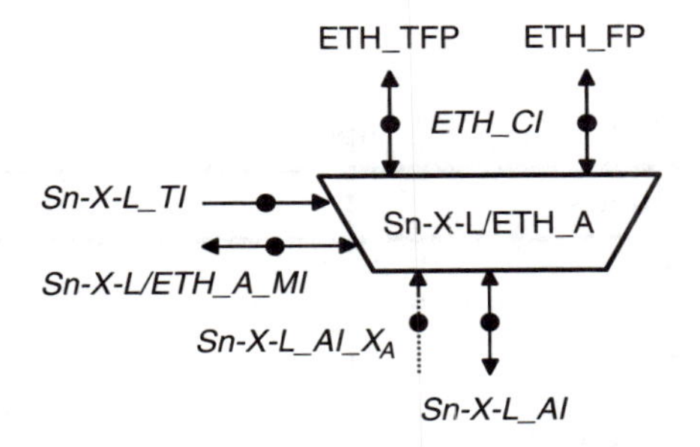

Figure 5.16 Bi-directional Sn-X-L/ETH_A_function.

The uni-directional representations of this function are depicted in Figures 5.17 and 5.19.

(11.1.2) LCAS-capable VC–n–Xv/ETH adaptation functions (Sn–X–L/ETH_A; n = (3, 4)

(11.1.2.1) LCAS-capable VC–n–Xv/ETH adaptation source function (Sn–X–L/ETH_A_So)

This function maps ETH_CI information onto an Sn–X–L_AI signal (n = 3 or 4).

The adapted information at the Sn–X–L_AP is a VC–n–X (n = 3 or 4), having a payload structure as described in ITU-T Rec. G.707 (2003), but with indeterminate POH bytes: J1, B3, G1. The information flow and processing of the Sn–X–L/ETH_A_So function are defined with reference to Figures 5.17 and 5.18.

Symbol

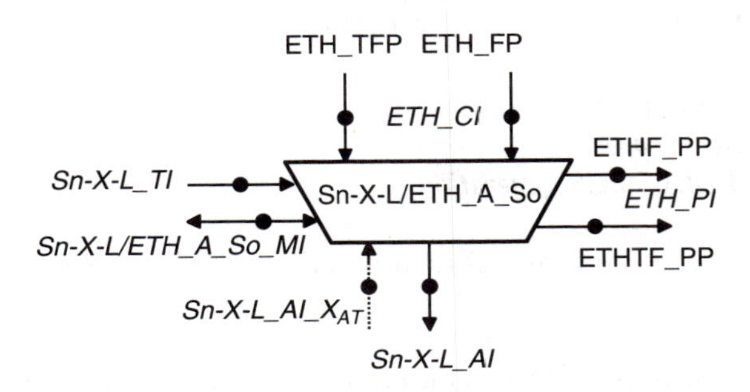

Figure 5.17 Sn-X-L/ETH_A_So function.

Interfaces

Sn–X–L/ETH_A_So input and output signals.

inputs	outputs
at the ETH_TFP: ETH_CI_Data **at the ETH_FP:** ETH_CI_Data ETH_CI_SSF **at the Sn–X–L_AP:** Sn–X–L_AI_X_{AT} **at the Sn_So_TP** Sn_TI_Clock Sn_TI_FrameStart **at the Sn/Sm_A_So_MP** Sn–X–L/ETH_A_So_MI_Active Sn–X–L/ETH_A_So_MI_CSFEnable	**at the Sn–X–L_So_AP** Sn_AI_Data Sn_AI_Clock Sn_AI_FrameStart **at the ETHF_PP:** ETH_PI_Data **at the ETHTF_PP:** ETH_PI_Data

Processes

A process diagram of this function is shown in Figure 5.18.

- ***Activation***: The Sn–X–L/ETH_A_So function shall access the access point when it is activated (MI_Active is true). Otherwise, it shall not access the access point. The next three processes are defined in ITU-T Rec. G.8021 (2004) clause 8.1.

  - *'Queuing' process*
  - *'Replicate' process*
  - *802.3 MAC FCS generation*

  The next three GFP processes are defined in ITU-T Rec. G.806 (2004) clause 8.5:

- ***Ethernet specific GFP-F source process***: In this function the GFP FCS generation is disabled (FCSenable = false). The UPI value

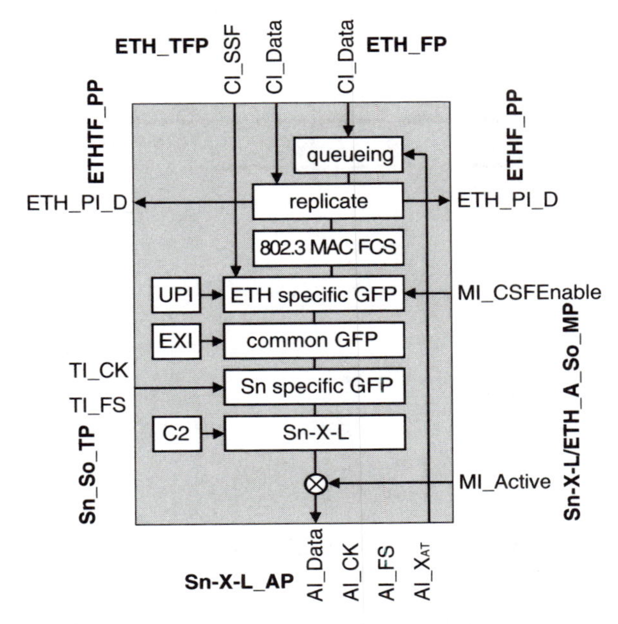

Figure 5.18 Sn-X-L/ETH_A_So processes.

'Frame-Mapped Ethernet' shall be inserted. The Ethernet frames are inserted into the client payload information field of the GFP-F frames according to ITU-T Rec. G.7041 (2003) clause 7.1. Response to ETH_CI_SSF asserted is *for further study*.

- ***Common GFP source process***: The GFP channel multiplexing is not supported (CMuxActive = false). The EXI value '0001' shall be inserted.
- ***VC–n specific GFP source process***: The GFP frames are mapped into the VC–n payload area according to ITU-T Rec. G.707 (2003) clause 10.6.
- ***VC–n specific source process***:

  - **C2**: Signal label information is derived directly from the adaptation function type. The value for 'GFP mapping' is placed in the C2 octet position.
  - **H4**: For the Sn–X–L/ETH_A_So function with n = 3, 4 the H4 octet is undefined at the Sn–X–L_AP output of this function (as well as the K3, F2, F3 octets); see ITU-T Rec. G.783 (2004) clause 12.

Defects

None.

Consequent actions

None.

Defect correlations

None.

Performance monitoring

For further study.

(11.1.2.2) LCAS-capable VC–n–Xv/ETH adaptation sink function (Sn–X–L/ETH_A_Sk)

This function extracts ETH_CI information from a VC–n-Xv server signal (n = 3 or 4), delivering ETH_CI to ETH_TFP and ETH_FP. The adapted information at the Sn–X–L_AP is a VC–n-Xv (n = 3 or 4), having a payload structure as described in ITU-T rec. G.707 (2003).

The information flow and processing of the Sn–X–L/ETH_A_Sk function are defined with reference to Figures 5.19 and 5.20.

Symbol

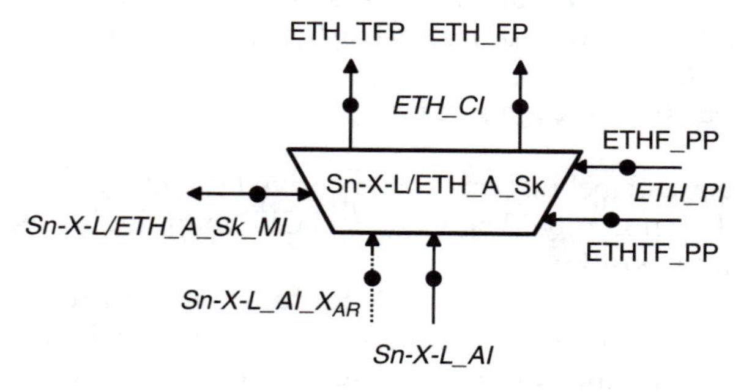

Figure 5.19 Sn-X-L/ETH_A_Sk function.

Interfaces

Sn–X–L/ETH_A_Sk input and output signals.

inputs	outputs
at the Sn–X–L_AP Sn–X–L_AI_Data Sn–X–L_AI_ClocK Sn–X–L_AI_FrameStart Sn–X–L_AI_TSF Sn–X–L_AI_X_{AR}	**at the ETH_TFP** ETH_CI_Data ETH_CI_SSF **at the ETH_FP**: ETH_CI_Data ETH_CI_SSF
at the ETHF_PP: ETH_PI_Data **at the ETHTF_PP**: ETH_PI_Data	
at the Sn–X–L/ETH_A_Sk_MP Sn–X–L/ETH_A_Sk_MI_Active Sn–X–L/ETH_A_Sk_MI_FilterConfig Sn–X–L/ETH_A_Sk_MI_CSF_Reported	**at the Sn–X–L/ETH_A_Sk_MP**: Sn–X–L/ETH_A_Sk_MI_AcSL Sn–X–L/ETH_A_Sk_MI_AcEXI Sn–X–L/ETH_A_Sk_MI_AcUPI Sn–X–L/ETH_A_Sk_MI_cPLM Sn–X–L/ETH_A_Sk_MI_cLFD Sn–X–L/ETH_A_Sk_MI_cUPM Sn–X–L/ETH_A_Sk_MI_cEXM Sn–X–L/ETH_A_Sk_MI_cCSF Sn–X–L/ETH_A_Sk_MI_pFCSError

Processes

In this function the signal Sn–X–L_AI_X_{AR} is not connected to any of the internal processes.

- ***Activation***: The Sn–X–L/ETH_A_Sk function shall access the access point and perform the common and specific processes operation specified below when it is activated (MI_Active is true). Otherwise, it shall activate the SSF signals at its outputs (TFP and FP) and not report its status via the management point. These three processes are defined in ITU-T Rec. G.8021 (2004) clause 8.1.

  - *'Filter' process*
  - *'Replicate' process*
  - *'802.3 MAC frame check' process*

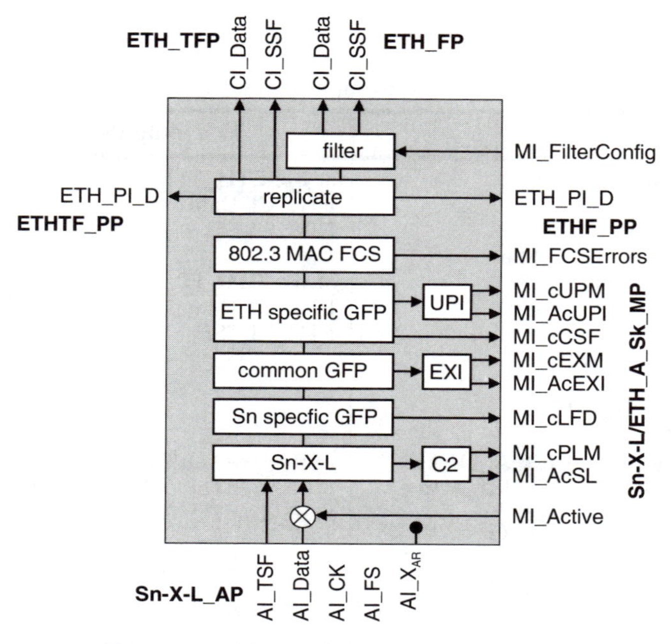

Figure 5.20 Sn-X-L/ETH_A_Sk processes.

The next three GFP processes are defined in ITU-T Rec. G.806 (2004) clause 8.5:

- ***Ethernet specific GFP-F sink process***: This function does not support GFP FCS checking (FCSdiscard = false), the performance primitives p_FCSError and p_FD is are not used. The UPI value 'Frame-Mapped Ethernet' shall be expected. The Ethernet frames are extracted from the client payload information field of the GFP-F frames according to ITU-T Rec. G.7041 (2003) clause 7.1. The accepted UPI value AcUPI is also available at the Sn–X–L/ ETH_A_Sk_MP.
- ***Common GFP sink process***: This function does not support GFP channel multiplexing (MI_CMuxActive = false). The EXI value '0001' shall be expected. The accepted EXI value AcEXI is also available at the Sn–X–L/ETH_A_Sk_MP.
- ***VC–n specific GFP sink process***: The GFP frames are demapped from the VC–n payload area according to ITU-T Rec. G.707 (2003) clause 10.6.

- ***VC–n specific sink process***:

  - **C2**: The signal label is recovered from the C2 byte as per ITU-T Rec. G.806 (2004) clause 6.2.4.2. The signal label value 'GFP mapping' shall be expected. The accepted value of the signal label AcSL is also available at the Sn–X–L/ETH_A_Sk_MP.

Defects

The defects dPLM, dLFD, dUPM and dEXM shall be detected in this function according to the specification in ITU-T Rec. 806 (2004) clause 6.2.

Consequent actions

The function shall perform the following consequent actions:

aSSF ← AI_TSF or dPLM or dLFD or dUPM or dEXM or dCSF

(Note: $X_{AR} = 0$ results in AI_TSF being asserted in the preceding Sn–X–L_TT function, so there is no need to include it as an additional contributor to aSSF.)

Defect correlations

The function shall perform the following defect correlations to determine the most probable fault cause as defined in ITU-T Rec. G.806 (2004) clause 6.4. These fault causes shall be reported to the element manger via the Sn–X–L/ETH_A_Sk_MP.

cPLM ← dPLM and (not AI_TSF)
cLFD ← dLFD and (not dPLM) and (not AI_TSF)
cUPM ← dUPM and (not dPLM) and (not dLFD) and (not AI_TSF)
cEXM ← dEXM and (not dUPM) and (not dPLM) and (not dLFD) and (not AI_TSF)
cCSF ← per ITU-T Rec. G.806 (2004) clause 8.5.4.1.2.

Performance monitoring

The function shall perform the following performance monitoring primitives processing. This performance monitoring primitive shall be reported to the element manger via the Sn–X–L/ETH_ A_Sk_MP.

pFCSError	count of FrameCheckSequenceErrors per second

(Note: this primitive is calculated by the MAC frame check process.)

5.2.4 *GFP mapping in the Sn–X/<client>_A function*

The most obvious mapping methodology that can be used to enable the support of data transport in existing synchronous networks is the *Generic Framing Procedure* (GFP) as defined in ITU-T Rec. G.7041 (2003). GFP mapping is a function of the adaptation function Sn–X/<client>_A where a packet based client signal is mapped into a *constant bitrate* (CBR) signal Sn–X and X is application specific, e.g. 1–256 for an SDH S4 structure. The adaptation process using GFP can be separated into three generic parts as shown in Figure 5.21:

- one or more instances of client layer specific processes with a client GFP part;
- common GFP processes; and
- server layer specific processes with a server GFP part.

This section only describes the GFP related functionality of the adaptation function.

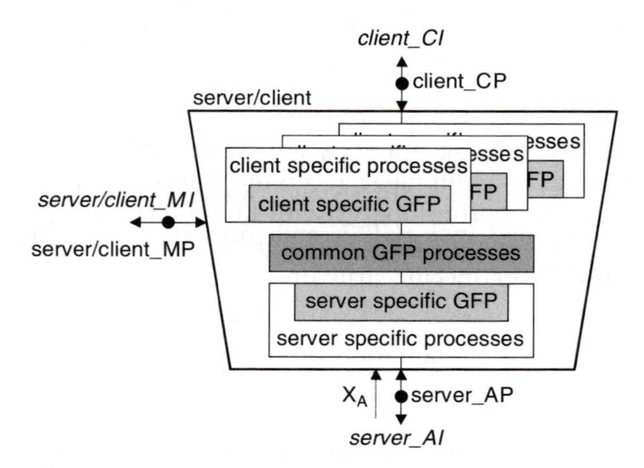

Figure 5.21 The server/client GFP adaptation function.

5.2.4.1 Source side GFP adaptation processes

Client specific GFP mapping

The client specific GFP processes perform the mapping of client data into GFP frames. There are two types of mapping: frame-mapped GFP (GFP-F) and transparent-mapped GFP (GFP-T). Client layer specific processes are defined in the adaptation functions of the technology specific equipment recommendations.

Client specific GFP-F source processes

The inputs to this process are client frames and optionally a Signal Fail (*Client_SF*). The received data are processed on a frame per frame base. The output is a CBR signal consisting of GFP frames.

Client specific GFP-T source processes

The input to this process is a stream of data and control words and a Loss of Signal (*Client_LOS*) and Loss of Character Synchronization (*Client_LCS*) indication from the client layer. The received data are processed on a character or block of characters base. The output is a CBR signal consisting of GFP frames.

Common GFP source processes

The processes are performed on a frame per frame base. The input and the output are CBR signals consisting of GFP frames.

Server layer specific GFP source processes

The input is a CBR signal consisting of GFP frames. Server layer specific processes are defined in the adaptation functions of the technology specific equipment recommendations. For example, in the application supporting VCAT with LCAS enabled the signal Sn–X–L_AI_X_{AT} shall control the GFP frame rate. The output is a signal with the server layer specific structure.

5.2.4.2 Sink side GFP adaptation processes

Client specific GFP de-mapping

The client specific GFP processes at the sink side perform the de-mapping of client data frames from GFP frames. Two types of de-mapping, GFP-F and GFP-T, are defined. Client layer specific processes are defined in the adaptation functions of the technology specific equipment recommendations.

Client specific GFP-F sink processes

The received signal is processed on a frame per frame base. The input consists of GFP frames; output is the original client frame and a possible Signal Fail (*Client_SF*).

Client specific GFP-T sink processes

The input consists of GFP frames; the output is a stream of client data and client control codeword octets.

Common GFP sink processes

The processes are performed on a frame per frame base. The input and output consist of GFP frames.

Server layer specific GFP sink processes

The input to the processes is the server layer data. Server layer specific processes are defined in the adaptation functions of the technology specific equipment recommendations. For example, in the application supporting VCAT with LCAS enabled the signal $Sn–X–L\_AI\_X_{AT}$ will control the GFP frame rate.

6

Trail termination functions

Or trail creation functions

This chapter describes the generic atomic model that represents the creation and/or termination of a trail in a layer network. This is generally referred to as the trail termination function or abbreviated to termination function. Even though the generic model is fairly simple, applying the expansion methodology described in Chapter 4 Section 4.2.2 can raise the level of detail in the atomic function. To clarify the application of trail termination functions some examples of existing trail termination models are provided.

6.1 GENERIC TRAIL TERMINATION FUNCTION

This atomic function is located within a layer network. The main purpose of this function is to monitor the integrity of the transported client signal and supervise its transport through the layer network. The client signal has been adapted for the transport by the appropriate adaptation function as described in Chapter 5 and is present at the access point of a termination function. The transport entity present between two access points is referred to as a trail. The information required for the monitoring and supervision of the adapted information may be generated and added at the ingress and extracted and analyzed at the egress. The termination source function at the ingress has the capability to create connectivity and performance monitoring information, generally referred to as overhead, without affecting the client or payload signal. The termination sink function at the egress has the capability to check the

SDH/SONET Explained in Functional Models Huub van Helvoort

connectivity and monitor the performance. The sink function will remove the overhead from the transported signal, i.e. the trail overhead is terminated. Depending on the technology of the layer network, the overhead can be provided by increasing the transported capacity or using already available, but unused, capacity.

Figure 6.1 shows the bi-directional model of a trail termination. In this generic model the designation <layer> has been used to identify the technology of the layer network. When the model is used in a specific technology the <layer> has to be replaced, for example, by Sn, MSn, ODUkP, EthP, etc.

In the following description of the trail termination function the same documentation template is used as the template used in most of the ETSI standards and ITU-T Recommendations. It contains separate paragraphs for:

- **symbol**, a figure representing the function in a functional model;
- **interfaces**, a table containing input and output signals;
- **processes**, detailed description of the function specific processes;
- **defects**, list of the defects detected by this function, e.g. dUNEQ;
- **consequent actions**, list of the actions based on the detected defects, e.g. aAIS;
- **defect correlation**, list of the correlations used to determine the most probable cause of defects, e.g. cSSF;
- **performance monitoring**, list of the performance monitoring primitives, e.g. pN_DS.

Bi-directional model: <layer>_TT

Symbol

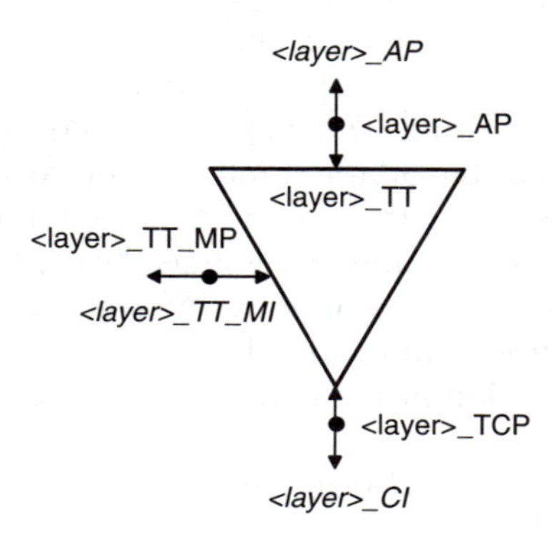

Figure 6.1 Bi-directional <layer>_TT function.

This atomic function can be decomposed into the generally co-located uni-directional trail termination source function and termination sink function. The models are illustrated in Figures 6.2 and 6.4.

Source function: <layer>_TT_So

The termination source function adds error monitoring, connectivity and status overhead information to the <layer> adapted information presented at its input by the server-to-client layer adaptation function to form the <layer> characteristic information that is presented at its output to be transported by the <layer> connection function to the termination sink function. The information flow and the processing of the <layer>_TT_So function are described with reference to Figures 6.2 and 6.3.

Symbol

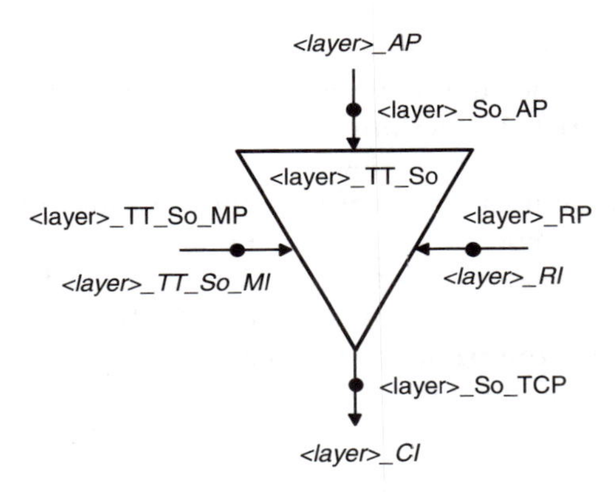

Figure 6.2 Source side <layer>_TT function.

Interfaces

<layer>_TT_So input and output signals.

inputs	outputs
at the <layer>_So_AP layer adapted information <layer>_AI_nn **at the <layer>TT_So_MP** layer TT management information <layer>_TT_So_MI_nn **at the <layer>_So_RP** layer remote information <layer>_RI_nn	**at the <layer>_So_TCP** layer characteristic information <layer>_CI_nn

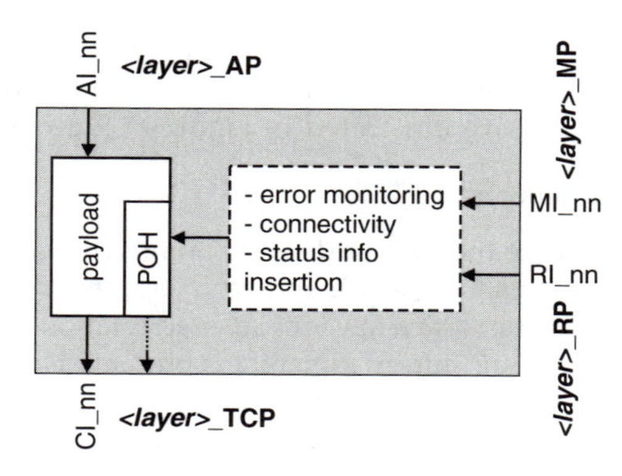

Figure 6.3 <layer>_TT_So processes.

Processes

The processes associated with the <layer>_TT_So function are as depicted in Figure 6.3. Typically, there are three processes:

- provisioning and insertion of identification information to enable checking of the connectivity in the network;
- generation and insertion of *bit-interleaved parity* (BIP), *cyclic redundancy check* (CRC) or *frame check sequence* (FCS) to enable checking of the integrity of the payload transport;
- returning to the far-end the status of the payload received by the associated sink function for analyzing the performance of the transport.

Defect

In general, no defects are detected by a source termination function.

Consequent actions

In general, no consequent actions are taken by a source termination function.

Defect correlation

Since there are no defects there is also no correlation of defects.

Performance monitoring

In general, there is no performance monitoring by a source termination function.

Sink function: <layer>_TT_Sk

The termination sink function analyzes and removes the overhead information of the <layer> characteristic information presented to its input by the <layer> connection function and presents the resulting client layer adapted information to its output for further processing by the <layer> to client layer adaptation function. The information flow and the processing of the <layer>_TT_Sk function are described with reference to Figures 6.4 and 6.5.

Symbol

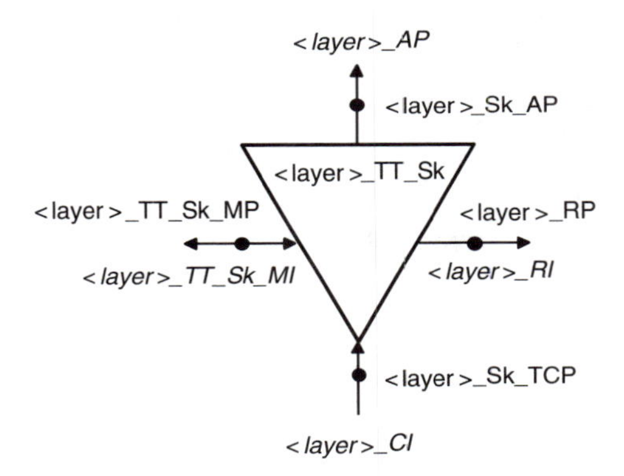

Figure 6.4 Sink side <layer>_TT function.

Interfaces

<layer>_TT_Sk input and output signals.

inputs	outputs
at the <layer>_Sk_TCP layer characteristic information <layer>_CI_nn **at the <layer>_TT_Sk_MP** layer TT management information <layer>_TT_Sk_MI_nn	**at the <layer>_Sk_AP** layer adapted information <layer>_AI_nn **at the <layer>_TT_Sk_MP** layer TT management information <layer>_TT_Sk_MI_nn **at the <layer>_RP** layer remote information <layer>_RI_nn

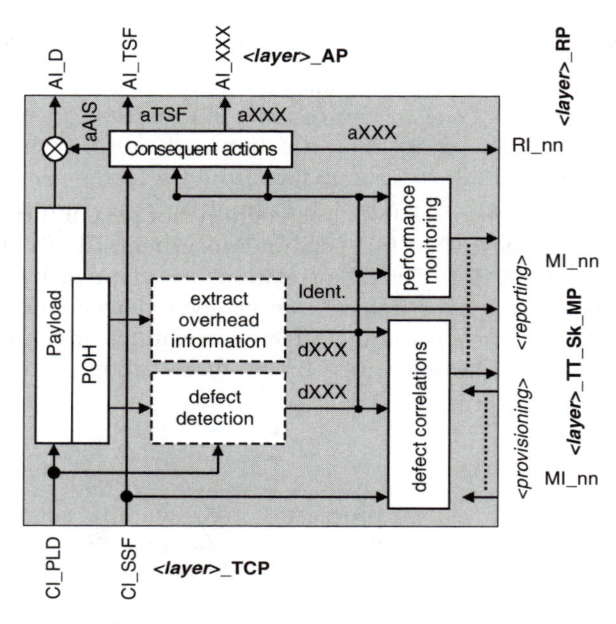

Figure 6.5 <layer>_TT_Sk processes.

Processes

The processes associated with the <layer>_TT_Sk function are as depicted in Figure 6.5. The error monitoring, identity and status overhead information is recovered to determine the status, connectivity and performance of the trail through the network.

Defects

Definitions of defect processing are generally not part of the functional specification but described in a separate document, e.g. in ITU-T Rec. G.806 (2004) clause 6.2. Defects are identified by an abbreviation prefixed with a 'd': dXXX, e.g. dTIM.

Consequent actions

Definitions of consequent actions are generally not part of the functional specification but described in a separate document, e.g. in ITU-T Rec. G.806 (2004) clause 6.3. Consequent actions are defined by equations containing detected layer defects (e.g. CI_SSF), server layer defects (e.g. dTIM) and can be disabled by provisioning (e.g. MI_TIMdis). Consequent actions are identified by an abbreviation prefixed with an 'a': aXXX, e.g. dTSF.

Defect correlation

Definitions of defect correlation are generally not part of the functional specification but described in a separate document, e.g. in ITU-T Rec. G.806 (2004) clause 6.4. Defect correlations are defined by equations containing detected layer defects, server layer defects and reporting can be disabled by provisioning. Defects correlations are used to find a probable cause and are identified by an abbreviation prefixed with a 'c': cXXX, e.g. cUNEQ.

Performance monitoring

Definitions of performance processing are generally not part of the functional specification but described in a separate document, e.g. in ITU-T Rec. G.806 (2004) clause 6.5. The number of errors and defects is counted per second. Performance is defined by equations containing detected layer defects, server layer defects or detected bit/block errors. Performance primitives are identified by an abbreviation prefixed with a 'p': pXXX, e.g. pN_DS.

6.2 TRAIL TERMINATION FUNCTION EXAMPLES

There are already many different trail termination functions defined, see ETSI standards EN300 417-1 to 417-7 and 417-9 to 417-10 and ITU-T Rec. G.705 (2000) for PDH, G.783 (2004) for SDH, G.798 (2004) for OTN, I.732 (2000) for ATM, G.8021 (2004) for Ethernet and G.mplseq (2005) for MPLS. In this subsection only a subset is taken and described to serve as an example for new developments.

6.2.1 *The Sn_TT function*

The first example is derived from ITU-T Rec. G.783 (2004) clause 12.2.1: the VC–n layer trail termination Sn_TT. In this technology the trail overhead is transported in-band, i.e. the transported signal structure contains octets dedicated to the overhead.

(12.2.1) VC–n layer trail termination Sn_TT

The Sn_TT_So function at the start of the trail creates a *virtual container* VC–n (n = 3, 4, 4–Xc) at the Sn_TCP by generating and adding *path overhead* POH information to the adapted information received at the Sn_AP. The Sn_TT_Sk function at the end of the trail terminates and processes the POH to determine the status of the defined path attributes. The structure of the adapted information, i.e. a synchronous nth order container C–n (n = 3, 4, 4–Xc), and the POH is defined in ITU-T Rec. G.707 (2003). The symbol of the bi-directional model is shown in Figure 6.6.

Symbol

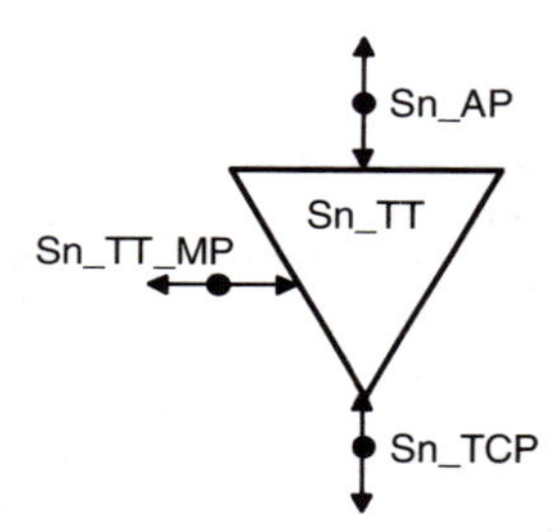

Figure 6.6 Bi-directional Sn_TT function.

This model can be further decomposed into the uni-directional source and sink models. These models are illustrates in Figures 6.7 and 6.9.

(12.2.1.1) VC–n layer trail termination source Sn_TT_So

The source function adds error monitoring and status overhead bytes to the adapted information received at the Sn_AP. The adapted information received at the Sn_AP is a VC–n (n = 3, 4, 4–Xc), having a payload as described in ITU-T Rec. G.707 (2003), but with indeterminate VC–n POH bytes: J1, B3, G1. These POH bytes are set as part of the Sn_TT function and the complete VC–n is forwarded to the Sn_TCP. The information flow and the processing of the Sn_TT_So function are defined with reference to Figures 6.7 and 6.8.

Symbol

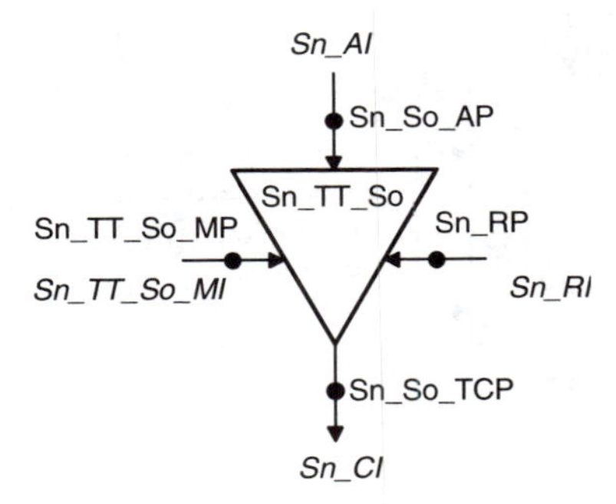

Figure 6.7 Source side Sn_TT function.

Interfaces

Sn_TT_So input and output signals.

inputs	outputs
at the Sn_So_AP Sn_AI_Data Sn_AI_ClocK Sn_AI_FrameStart **at the Sn_TT_So_MP** Sn_TT_So_MI_TxTI **at the Sn_So_RP** Sn_RI_RDI Sn_RI_REI	**at the Sn_So_TCP** Sn_CI_Data Sn_CI_ClocK Sn_CI_FrameStart

Processes

The processes associated with the Sn_TT_So function are as depicted in Figure 6.8.

- **J1**: The *trail trace identifier* (TTI) should be generated and placed in the TTI octet position of the POH. Its value is provisioned via the reference point Sn_TT_So_MP. The format of the TTI is described in ITU-T Rec. G.806 (2004) clause 6.2.2.2.
- **B3**: The *bit interleaved parity* (BIP-8) is computed over all bits of the previous VC–n frame and placed in B3 octet position of the POH.
- **G1**: The performance monitoring octet. The number of remote errors Sn_RI_REI detected and sent by the associated sink function is encoded in the REI position, i.e. bits 1 to 4, of the G1 octet of the

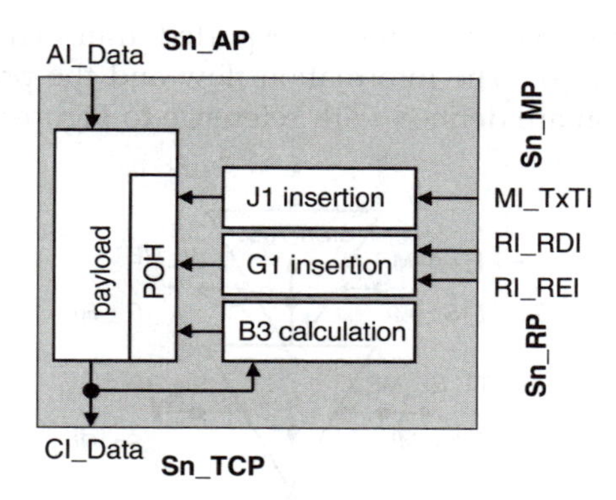

Figure 6.8 Sn_TT_So process.

POH. The REI information shall be encoded within 1 ms after the detection of the errors.

The remote defect Sn_RI_RDI detected and sent by the sink function shall be inserted in the RDI position, i.e. bit 5, of the G1 octet in the POH. The RDI information shall be inserted within 1 ms after the detection or clearing of the defect. Bits 6 and 7 of the G1 octet are reserved for the optional use of enhanced-RDI a SONET specific defect. The default value is 00 or 11.

Defects

None.

Consequent actions

None.

Defect correlations

None.

Performance monitoring

None.

(12.2.1.2) VC–n layer trail termination sink Sn_TT_Sk

The sink function monitors the VC–n (n = (3, 4, 4–Xc)) structure received at the Sn_TCP for errors, and recovers the trail status from the POH. It extracts the payload-independent overhead octets J1, B3 and G1 from the VC–n layer characteristic information to verify the connectivity and monitor the performance of the transported trail. The VC–n is forwarded to the Sn_AP. The information flow and the processing of the Sn_TT_Sk function are defined with reference to Figures 6.9 and 6.10.

Symbol

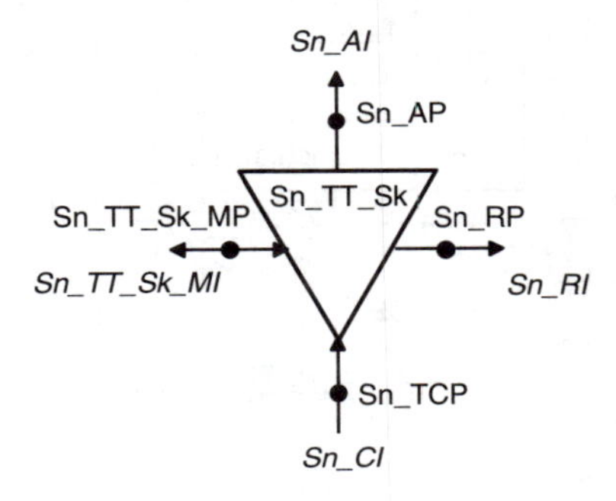

Figure 6.9 Sink side Sn_TT function.

Interfaces

Sn_TT_Sk input and output signals.

inputs	outputs
at the Sn_So_TCP	**at the Sn_So_AP**
Sn–X_CI_Data	Sn_AI_Data
Sn–X_CI_ClocK	Sn_AI_ClocK
Sn–X_CI_FrameStart	Sn_AI_FrameStart
Sn–X_CI_SSF	Sn_AI_TSF
	Sn_AI_TSD
at the Sn_TT_Sk_MP	**at the Sn_TT_Sk_MP**
Sn_TT_Sk_MI_TPmode	Sn_TT_Sk_MI_AcTI
Sn_TT_Sk_MI_ExTI	Sn_TT_Sk_MI_cTIM
Sn_TT_Sk_MI_RDI_Reported	Sn_TT_Sk_MI_cUNEQ
Sn_TT_Sk_MI_SSF_Reported	Sn_TT_Sk_MI_cEXC
Sn_TT_Sk_MI_DEGTHR	Sn_TT_Sk_MI_cDEG
Sn_TT_Sk_MI_DEGM	Sn_TT_Sk_MI_cRDI
Sn_TT_Sk_MI_EXC_X	Sn_TT_Sk_MI_cSSF
Sn_TT_Sk_MI_DEG_X	Sn_TT_Sk_MI_pN_EBC
Sn_TT_Sk_MI_1second	Sn_TT_Sk_MI_pF_EBC
Sn_TT_Sk_MI_TIMdis	Sn_TT_Sk_MI_pN_DS
Sn_TT_Sk_MI_TIMAISdis	Sn_TT_Sk_MI_pF_DS
	at the Sn_Sk_RP
	Sn_RI_RDI
	Sn_RI_REI

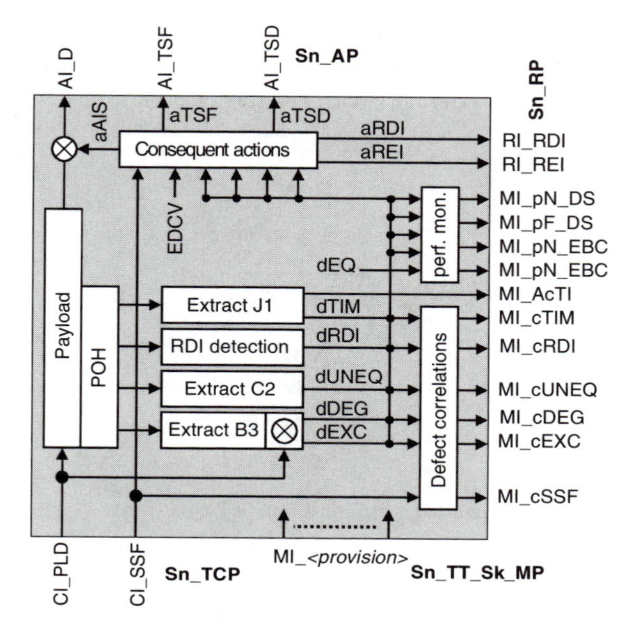

Figure 6.10 Sn_TT_Sk processes.

Processes

The processes associated with the Sn_TT_Sk function are as depicted in Figure 6.10.

- **J1**: The TTI is recovered from VC–n POH received at the Sn_TCP and processed as specified in ITU-T Rec. G.806 (2004) clause 6.2.2.2. The expected value of the TTI, ExTI, and the accepted value of the TTI, AcTI, are used to detect a possible mismatch defect cTIM. TIMdis can disable the reporting of cTIM and TIMAISdis can disable the consequent AIS insertion.
- **C2**: The signal label; an unequipped defect cUNEQ shall be detected.
- **B3**: The BIP-8 is computed for each VC–n frame. The computed BIP-8 value of the current frame is compared to the B3 octet in the POH of the following frame to determine the number of errors. The process shall detect excessive errors cEXC and signal degrade cDEG. Input parameters for this process are: EXC_X, DEG_X, DEG_M, DEGTHR.

- **G1**: The REI information shall be recovered from the G1 octet bits 1–4 and the derived performance primitives should be reported at the Sn_TT_Sk_MP. Bit 5 in the G1 octet is processed to detect the cRDI defect. The performance of the trail can be derived from the REI and RDI information and will be reported as near-end defect seconds pN_DS, far-end defect seconds pF_DS, near-end error block count pN_EBC and far-end error block count pF_EBC. To enable correlation of performance data in a domain, a one second timestamp '1 second' is distributed to all performance processes. Bits 6 and 7 of the G1 octet shall be ignored if the option enhanced-RDI is not used.
- **N1**: The network operator octet N1 is defined for tandem connection monitoring TCM. TCM is described in Chapter 8, Section 8.3.2. The TCM specific termination function is described in ITU-T Rec. G.783 (2004) clauses 12.4.2 and 12.4.3.
- **K3**: Bits 5–8 are undefined and shall be ignored by this function.

If the Termination Point mode (TPmode) is true all defects and performance data will be reported as management information (Sn_TT_Sk_MI) to the element management function.

Defects

The defects dUNEQ, dTIM, dEXC, dDEG and dRDI shall be detected in this function according to the specification in ITU-T Rec. 806 (2004) clause 6.2. The defect dEQ represents a hardware, i.e. equipment, error that affects the transported signal.

Consequent actions

The function shall perform the following consequent actions as described in ITU-T Rec. G.806 (2004) clause 6.3.

aAIS ← dUNEQ or (dTIM and not TIMAISdis)
aRDI ← CI_SSF or dUNEQ or dTIM
aREI ← 'number of error detection code violations EDCV'
aTSF ← CI_SSF or dUNEQ or (dTIM and not TIMAISdis)
aTSFprot ← aTSF or dEXC
aTSD ← dDEG

On declaration of aAIS, the function shall output an all-ONEs (AIS) signal complying with the frequency limits for this signal within two VC–n frames (250 μs). Upon termination of the above failure conditions, the all-ONEs shall be removed within two VC–n frames (250 μs). aRDI and aREI are reported to the far-end as remote information Sn_RI to the associated source function.

Defect correlation

The function shall perform the following defect correlations to determine the most probable fault cause as defined in ITU-T Rec. G.806 (2004) clause 6.4. This fault cause shall be reported to the element manger via the Sn_TT_Sk_MP.

$$
\begin{aligned}
\text{cSSF} &\leftarrow \text{CI\_SSF and SSF\_Reported and MON}\\
\text{cUNEQ} &\leftarrow \text{dUNEQ and MON}\\
\text{cTIM} &\leftarrow \text{dTIM and (not dUNEQ) and MON}\\
\text{cEXC} &\leftarrow \text{dEXC and (not dTIM or TIMAISdis) and MON}\\
\text{cDEG} &\leftarrow \text{dDEG and (not dTIM or TIMAISdis) and MON}\\
\text{cRDI} &\leftarrow \text{dRDI and (not dUNEQ) and (not dTIM or TIMAISdis)}\\
&\quad\ \text{and MON and RDI\_Reported}
\end{aligned}
$$

Performance monitoring

The function shall perform the following performance monitoring primitives processing see ITU-T Rec. G.806 (2004) clause 6.5. The performance monitoring primitives shall be reported to the element management function.

$$
\begin{aligned}
\text{pN\_DS} &\leftarrow \text{CI\_SSF or dUNEQ or dTIM or dEQ}\\
\text{pF\_DS} &\leftarrow \text{dRDI}\\
\text{pN\_EBC} &\leftarrow \Sigma\text{nN\_B}\\
\text{pF\_EBC} &\leftarrow \Sigma\text{nF\_B}
\end{aligned}
$$

6.2.2 *The OCh_TT function*

The second example is derived from ITU-T Rec. G.798 (2004) clause 12.2.1: the OCh layer trail termination OCh_TT. In this technology the trail overhead is transported out-of-band, i.e. in an Optical Transport Module OTM Overhead Signal OOS.

(12.2.1) OCh trail termination function (OCh_TT)

The OCh_TT functions are responsible for the end-to-end supervision of the OCh trail. They provide full functionality based on the non-associated overhead information. Figure 6.11 shows the combination of the uni-directional sink and source functions to form a bi-directional function.

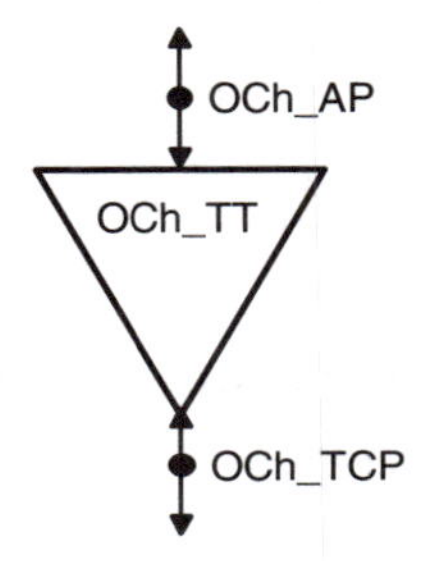

Figure 6.11 Bi-directional Och_TT function.

This model can be further decomposed into the uni-directional source and sink models. These models are illustrated in Figures 6.12 and 6.14.

(12.2.1.1) OCh trail termination source function (OCh_TT_So)

The OCh_TT_So function conditions the data received at the OCh_AP for transmission over the optical medium and presents them at the OCh_TCP. The information flow and the processing of the OCh_TT_So function are defined with reference to Figures 6.12 and 6.13.

Symbol

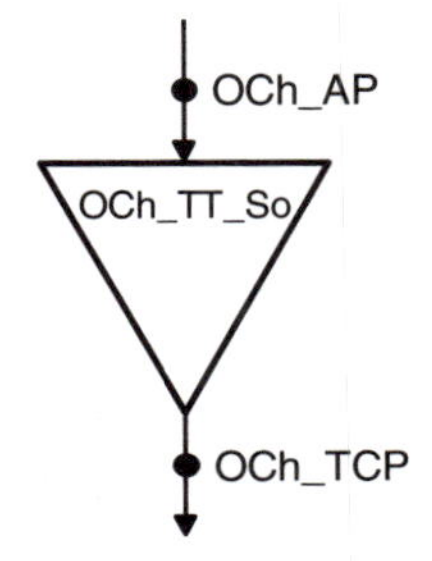

Figure 6.12 Och_TT_So function.

Interfaces

Och_TT_So input and output signals.

input	output
at the OCh_AP: OCh_AI_D	**at the OCh_TCP:** OCh_CI_PLD

Processes

The processes associated with the OCh_TT_So function are as depicted in Figure 6.13.

- Payload generation: The function shall generate the OCh payload signal (baseband signal).

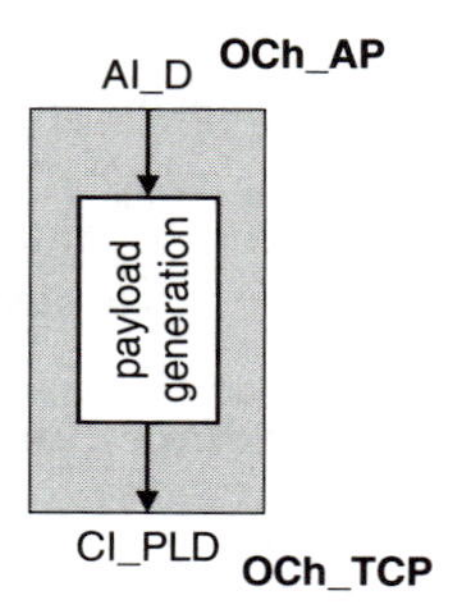

Figure 6.13 Och_TT_So process.

Defects

None.

Consequent actions

None.

Defect correlations

None.

Performance monitoring

None.

(12.2.1.2) OCh trail termination sink function (OCh_TT_Sk)

The OCh_TT_Sk function recovers the OCh payload signal and reports the performance of the OCh trail. It extracts the OCh overhead (including the, FDI-P, FDI-O, and OCI signals) from the OCh signal at its OCh_TCP and detects for LOS, OCI, FDI-P and FDI-O defects. The information flow and the processing of the OCh_TT_Sk function are defined with reference to Figures 6.14 and 6.15.

Symbol

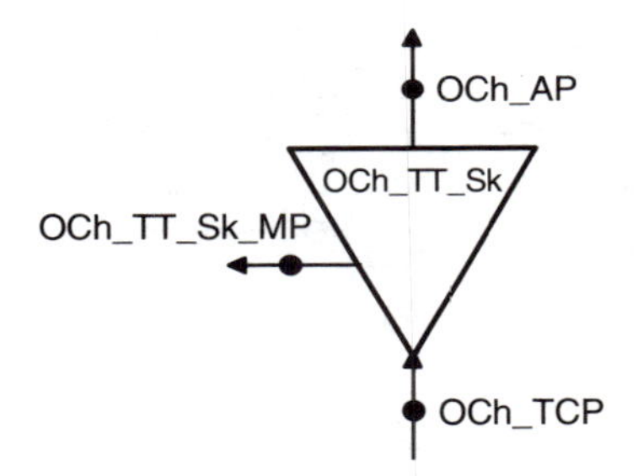

Figure 6.14 Och_TT_Sk function.

Interfaces

Och_TT_Sk input and output.

inputs	outputs
at the OCh_TCP: OCh_CI_PLD OCh_CI_OH OCh_CI_SSF-P OCh_CI_SSF-O	**at the OCh_AP:** OCh_AI_D OCh_AI_TSF-P OCh_AI_TSF-O **at the OCh_TT_Sk_MP:** OCh_TT_Sk_MI_cLOS-P OCh_TT_Sk_MI_cOCI OCh_TT_Sk_MI_cSSF OCh_TT_Sk_MI_cSSF-P OCh_TT_Sk_MI_cSSF-O

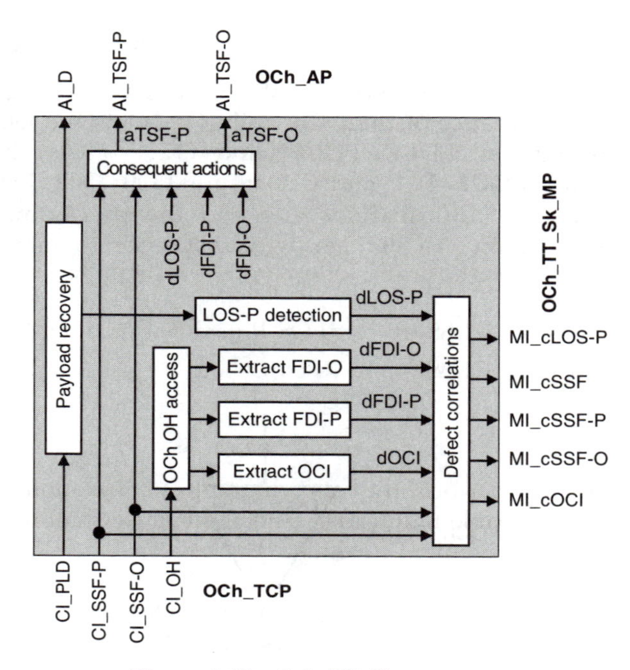

Figure 6.15 Och_TT_Sk processes.

Processes

The processes associated with the OCh_TT_So function are as depicted in Figure 6.15.

- **Payload recovery**: This function shall recover the OCh payload signal.
- **LOS-P**: The loss of signal; payload (OCh-LOS-P) shall be detected.
- **FDI-P**: The forward defect indication; payload information (OCh-FDI-P) shall be extracted from the OCh overhead of the OOS. It shall be used for FDI-P defect detection.
- **FDI-O**: The forward defect indication; overhead information (OCh-FDI-O) shall be extracted from the OCh overhead of the OOS. It shall be used for FDI-O defect detection.
- **OCI**: The open connection indication information (OCh-OCI) shall be extracted from the OCh overhead of the OOS. It shall be used for OCI defect detection.

Defects

The function shall detect dLOS-P, dFDI-P, dFDI-O and dOCI defects as defined in ITU-T Rec. G.798 (2004) clause 6.2. The defect dOCI shall be set to false during CI_SSF-O and dFDI-O.

Consequent actions

The function shall perform the following consequent actions:

aTSF-P ← CI_SSF-P or dLOS-P or dOCI or dFDI-P
aTSF-O ← CI_SSF-O or dFDI-O

Defect correlations

The function shall perform the following defect correlations to determine the most probable fault cause. This fault cause shall be reported to the element management function.

cLOS-P ← dLOS-P and (not dOCI) and (not FDI-P) and (not CI_SSF-P)
cOCI ← dOCI and (not CI_SSF-P) and (not CI_SSF-O)
and (not FDI-O) and (not FDI-P)
cSSF ← (CI_SSF-P or dFDI-P) and (CI_SSF-O or dFDI-O)
cSSF-P ← (CI_SSF-P or dFDI-P) and (not cSSF)
cSSF-O ← (CI_SSF-O or dFDI-O) and (not cSSF)

Performance monitoring

For further study.

6.2.3 *The ETH_FT function*

The third example could have been derived from ITU-T Rec. G.8021 (2004) clause 9.2. However, at the time of writing this book (April 2005), this clause indicated 'for further study'. Hence, the example provided is preliminary and not complete.

This example describes the Ethernet MAC layer flow termination function ETH_FT. In this technology the flow overhead is transported partly in-band and partly out-of-band.

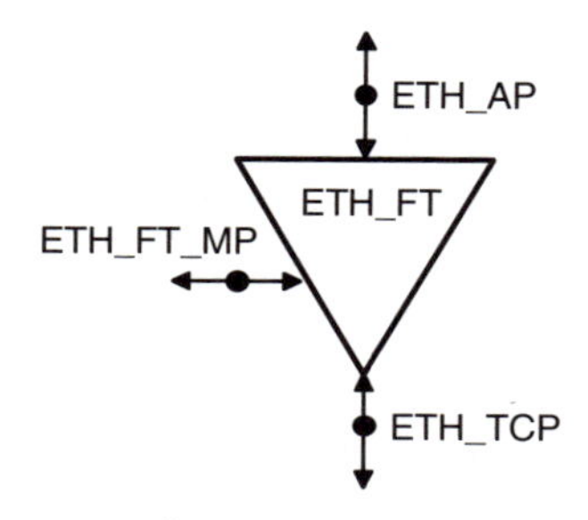

Figure 6.16 Bi-directional ETH_FT function.

(9.2) ETH flow termination function ETH_FT

The ETH_FT functions are responsible for the end-to-end supervision of the ETH flow. They provide full functionality based on the overhead information. Figure 6.16 shows the bi-directional function. The ETH layer network Characteristic Information (ETH_CI) is a (non-)continuous flow of ETH_CI traffic units. The ETH_CI traffic unit consists of the following set of signals: *Destination Address* (DA), *Source Address* (SA) and MAC *Service Data Unit* (M_SDU) (see ITU-T Rec. G.8010 (2004) and IEEE 802.3ae (2002) clause 2). Optionally, the M_SDU may include a *Priority Tag* (P) (see IEEE 802.1Q (2002). The ETH_CI traffic unit is transported over an ETH FPP link within a link specific frame or packet. (Note: The Preamble (PA), Start-of-Frame Delimiter (SFD) and 32-bit Frame Check Sequence (FCS) are considered part of the MAC frame (see IEEE 802.3 (2002) clause 3). In the layer network model, this PA/SFD/FCS is associated with the ETH FPP link, not with the ETH characteristic information. This modeling does not change the requirement, in IEEE 802.1D and IEEE 802.1Q, regarding introducing undetected frame errors.)

This model can be further decomposed into the uni-directional source and sink models. These models are illustrated in Figures 6.17 and 6.19.

(9.2.1) Eth trail termination source function (ETH_FT_So)

The ETH_FT_So function conditions the data received at the ETH_AP for transmission over the server layer network and presents them at the ETH_TFP. The information flow and the processing of the ETH_FT_So function are defined with reference to Figures 6.17 and 6.18.

Symbol

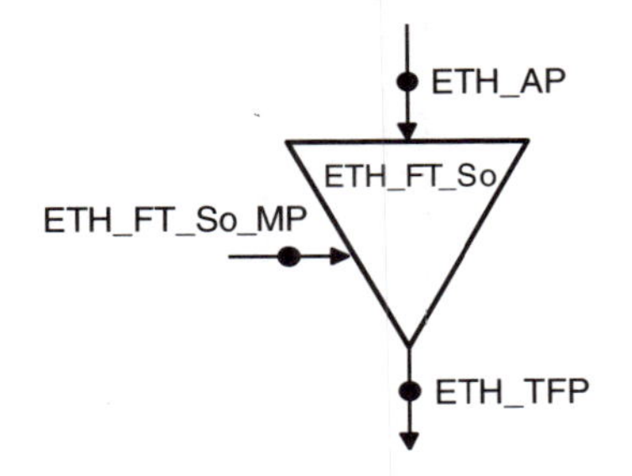

Figure 6.17 ETH_FT_ So function.

Interfaces

ETH_FT_So input and output.

input	output
at the ETH_AP: ETH_AI_Data ETH_AI_ClocK **at the ETH_FT_So_MP:** *for further study*	**at the ETH_TFP:** ETH_CI_Data ETH_CI_ClocK

Processes

The processes associated with the ETH_FT_So function are as depicted in Figure 6.18.

- **Payload encapsulation**: The function shall use the MAC protocol, defined in IEEE 802.3 (2002), to generate the ETH flow. This protocol encapsulates an SDU (payload data) by adding a 14-octet header, i.e. the *Protocol Control Information* (PCI) (DA/SA/P), in front of the payload data octets. If the ETH FPP link uses the MAC FCS to check the integrity of the transport, the additional FCS of each ETH_CI traffic unit is generated and appended as specified in IEEE 802.3 (2002) subclause 4.2.4.1.2

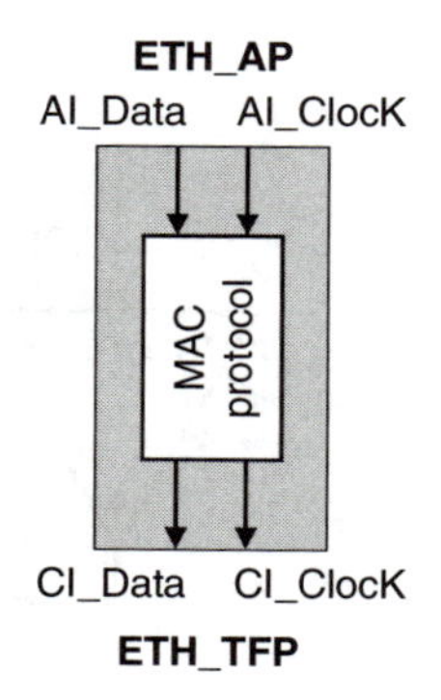

Figure 6.18 ETH_FT_So process.

- **OAM messages**: Further supervisory information may be sent from source to sink function by OAM messages. This is for further study.

Defects

None.

Consequent actions

None.

Defect correlations

None.

Performance monitoring

None.

(9.2.2) ETH trail termination sink function (ETH_FT_Sk)

The ETH_FT_Sk function recovers the payload signal and reports the performance of the ETH flow based on in-band and possible out-of-band information. The information flow and the processing of the ETH_FT_Sk function are defined with reference to Figures 6.19 and 6.20.

Symbol

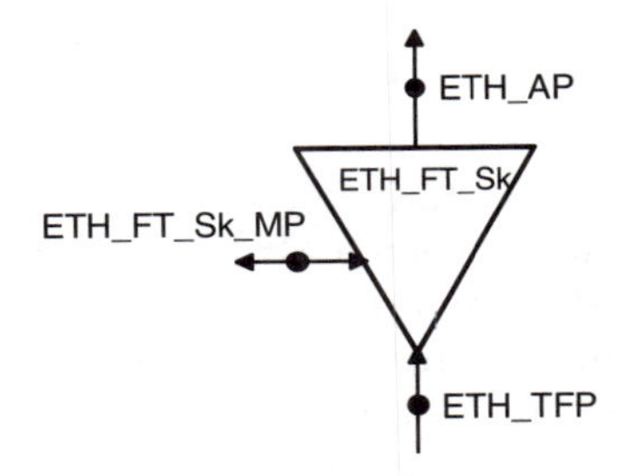

Figure 6.19 ETH_FT_Sk function.

Interfaces

ETH_FT_Sk input and output signals.

inputs	outputs
at the ETH_TFP: ETH_CI_Data ETH_CI_ClocK ETH_CI_SSF **at the ETH_FT_Sk_MP:** *for further study*	**at the ETH_AP:** ETH_AI_Data ETH_AI_ClocK ETH_AI_TSF **at the ETH_FT_Sk_MP:** ETH_FT_Sk_MI_cSSF ETH_FT_Sk_MI_FCSErrors *for further study*

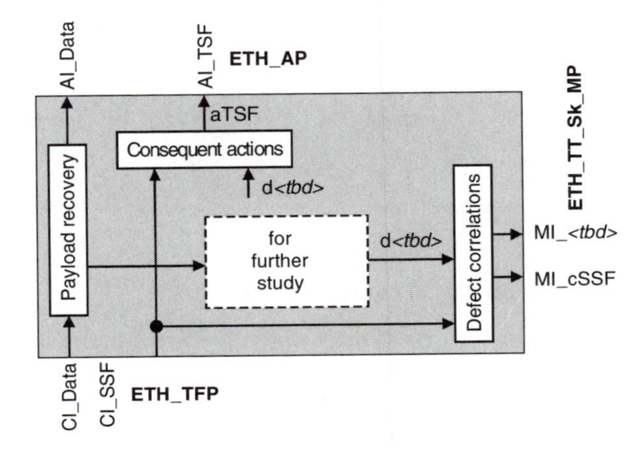

Figure 6.20 ETH_FT_Sk processes.

Processes

The processes associated with the ETH_FT_So function are as depicted in Figure 6.20.

- **Payload recovery**: This function shall recover the SDU payload signal using the MAC protocol. If the ETH FPP link includes the MAC FCS, the additional FCS of each ETH_CI traffic unit is checked as specified in IEEE 802.3 (2002) subclause 4.2.4.1.2. If errors are detected, the frame can be discarded. Errored frames may be reported by MI_FCSErrors.
- **OAM messages**: For further study.

Defects

For further study.

Consequent actions

For further study.

Defect correlations

For further study.

Performance monitoring

For further study.

7

Connection functions

The heart of the layer network

This chapter describes the generic atomic model that represents the connectivity in a layer network; it is generally referred to as the connection function (or flow domain in connectionless applications). Although the generic model is fairly simple the level of detail in the atomic function can be raised by applying the partitioning methodology described in Chapter 3, Section 3.2. An example of an existing connection model is provided to exercise the application of connection functions alongwith some specific applications of matrices, the most atomic form of a connection function.

7.1 GENERIC CONNECTION FUNCTION

This is the function within a connection-oriented layer network that provides the capability to transfer a signal with layer specific characteristic information received at one of its inputs to one or more of its outputs. In a connectionless layer network the term connection function shall be replaced by flow domain. The association between input and outputs, i.e. connections or flows, may be provisioned from an element or network management system. The connection function does not alter the information transferred from input to output. However, it may terminate any switching protocol information and act upon it. A connection function only has connection points to be bound with other atomic functions. If there are any (technology dependent) restrictions to

SDH/SONET Explained in Functional Models Huub van Helvoort

the connectivity, they shall be defined. In the following figures <layer> represents the notation used for the layer characteristic information <layer>_CI, e.g. Sn_CI, OTUk_CI, ODUk_ CI, ETH_CI, etc.

The standard convention to identify a connection function is <layer>_C. However, most of the time it is identified by just <layer> because the symbol identifies that it is a connection function. Note that a connection function is topologically identical to a sub-network and has the same properties. Figure 7.1 shows the bi-directional functional model of a connection function. Generally, flow domains are depicted as uni-directional models because of the uni-directional nature of flows.

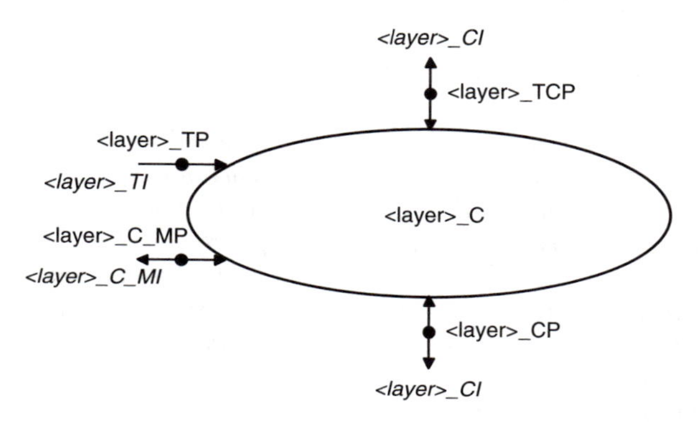

Figure 7.1 Bi-directional <layer>_C function.

Bi-directional model: <layer>_C

The connection point <layer>_TCP represents the binding with any of the layer trail termination functions <layer>_TT described in Chapter 5 and, only for the SDH S4 layer, the binding with the interworking function described in Chapter 10, Section 10.4. The connection point <layer>_CP represents the binding with an adaptation function <server>/<layer>_A adapting the characteristic information of this layer to the server layer as described in Chapter 6. In general, the CI will consist of the data signal together with the associated clock and framestart signals.

The connection management point <layer>_C_MP represents the binding with the management system and is used to provision the connection function and receive or retrieve information from the function.

The timing point <layer>_TP represents the binding with the timing layer and is used to convey *Timing Information* (TI) such as clock and frame synchronization signals.

The connection function can also be depicted as a uni-directional model as shown in Figure 7.2.

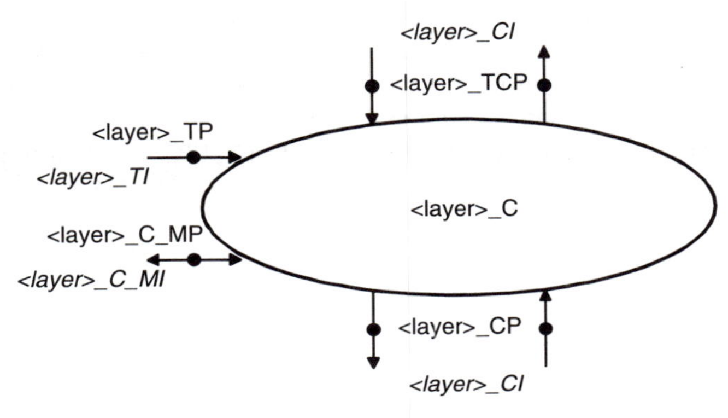

Figure 7.2 Uni-directiona l<layer>_C function.

Uni-directional: <layer>_C

This is the uni-directional model that provides the connection of a signal with characteristic information <layer>_CI received at one of its inputs to one or more of its outputs. The only difference with the bi-directional model is that it shows explicitly the inputs and outputs. The number of inputs and outputs towards the client layer and/or the server layer are not necessarily the same.

Interfaces

Table 7.1 <layer>_C input and output signals.

inputs	outputs
at <layer>_TCP layer signal characteristic information <layer>_CI_nn	**at <layer> TCP** layer signal characteristic information <layer>_CI_nn
at <layer>_CP layer signal characteristic information <layer>_CI_nn	**at <layer>_CP** layer signal characteristic information <layer>_CI_nn
at <layer>_C_MP connection management information <layer>_C_MI_nn	**at <layer>_C_MP** connection management information <layer>_C_MI_nn
at <layer>_TP connection timing information <layer>_TI_nn	

Processes

The function shall be able to connect any input with any output by means of establishing a matrix connection between a specified input and one or more specific outputs. It shall also be able to remove an established matrix connection. The network operator can provision matrix connections via the management interface. When the connection function supports protection switching, the protection process itself can also change the matrix connections. This is described in Chapter 9.

Routing

Each connection in the matrix can be provisioned by using the attributes shown in Table 7.2.

Table 7.2 Connection provisioning.

parameter	value
connection ports	input/output (T)CP identifiers
connection direction	– uni-directional – bi-directional
connection type	– unprotected – protected (type of protection)

Protection

It is possible to use the connectivity provided by the matrix in the protection schemes described in Chapter 9. However, this requires a protection switching process that supports one or more of these protection schemes and specific adaptation and trail termination functions in the same layer or in the server layer to provide the triggers for the protection switch.

Defects

Not specified.

Consequent actions

On every output port this function will send a signal with the appropriate <layer>_CI structure even when there is no connection provisioned to any of the input ports. It is recommended, in the characteristic information, to send an indication that the output port is not connected; this is technology dependent. For example, in SDH and OTN the signal is labeled as 'unequipped' and can be identified by a structure where the signal label and path trace octets have an all-zero value. This 'unequipped' signal can be used for fault location purposes, e.g. identify misconnections.

Defect correlation

Not specified.

Performance monitoring

Not specified.

7.2 CONNECTION FUNCTION EXAMPLE

This section provides two examples of the connectivity capability in layer networks:

- the connection-oriented VC–n layer connection function Sn_C;
- the connectionless Ethernet flow domain ETH_FD.

7.2.1 VC–n layer connection function Sn_C

The SDH equipment recommendations ITU-T Rec. G.783 (2004) contain several connection functions for different layer networks. This example shows only one of them: Sn_C defined in clause 12.1.1.

This is the function that transfers the VC–n signals present at its input ports to the VC–n signals at its output ports without altering their structure. (Note: Neither the number of input and output ports of the connection function, nor its connectivity are specified in any of the ETSI standards or ITU-T Recommendations. These are properties of individual network elements and are implementation dependent.)

The Sn_C connection process is a uni-directional function as illustrated in Figures 7.3 and 7.4. The signal structure at the input and output ports of the function are similar, differing only in the logical sequence of the VC–ns.

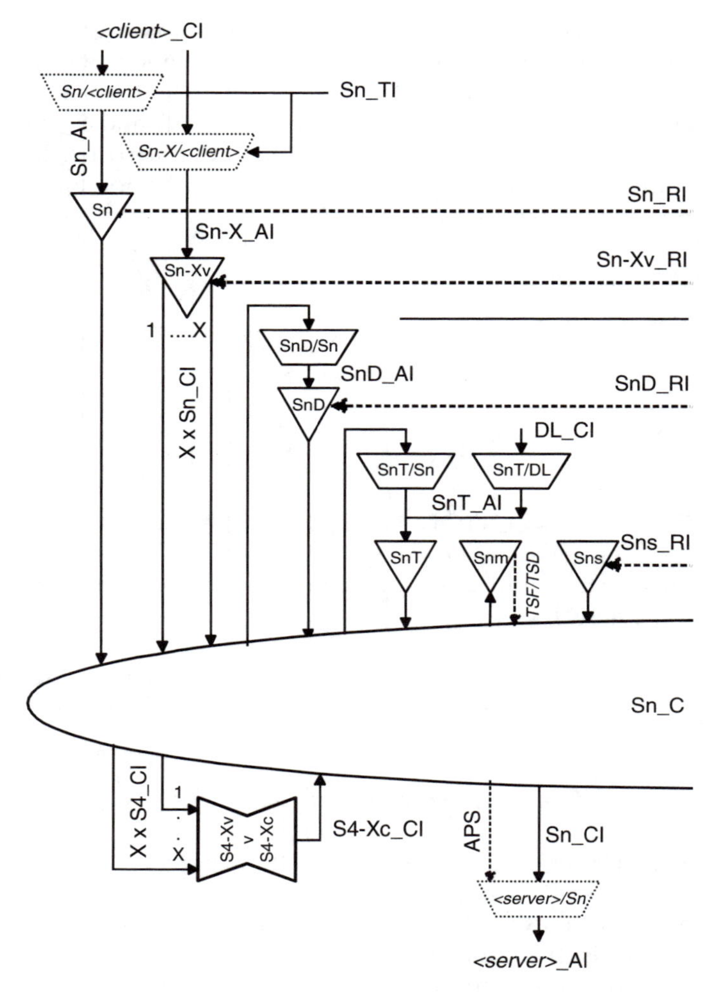

Figure 7.3a Sn layer network example; source side

As the process does not affect the structure of the characteristic information, the reference points at either side of the Sn_C function are the same, as shown in Figure 7.3. The reference points CP and TCP are identical; the latter identifies a specific binding with a trail termination function. Figure 7.3 presents a set of atomic functions

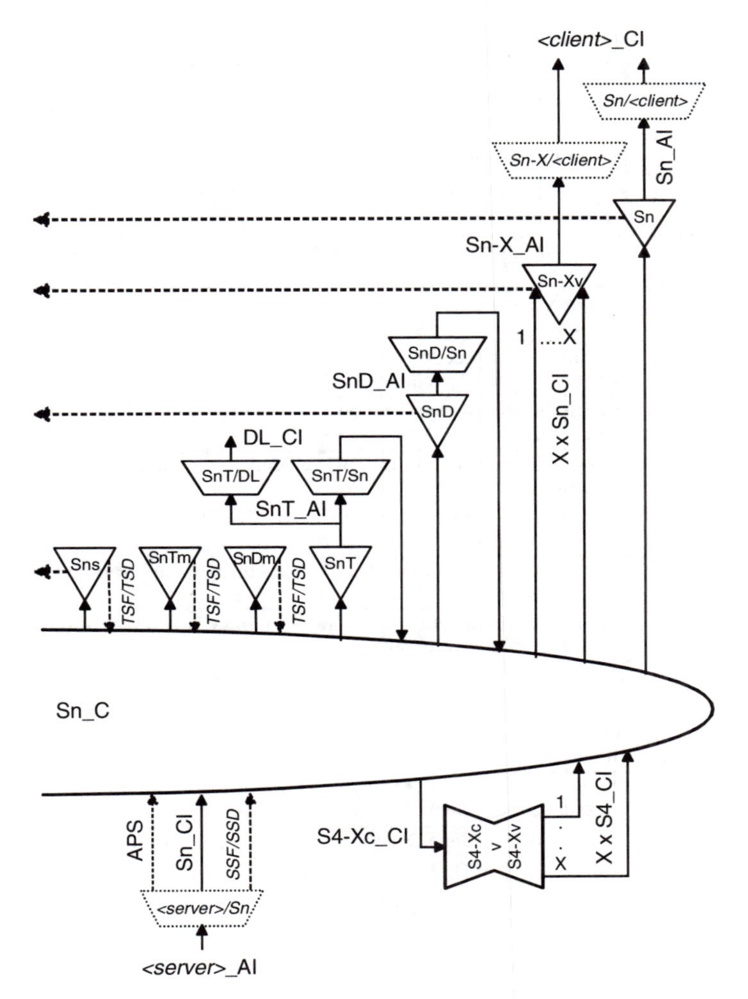

Figure 7.3b Sn layer network example; sink side.

that can be connected to this connection function in the VC–n layer network.

For a detailed description of the following atomic functions, see ITU-T Rec. G.783 (2004).

Table 7.3 Atomic functions that may be connected to an Sn_C.

function	description
Sn_C	VC–n layer connection function
Sn_TT	Sn trail termination source and sink functions
Sn–Xv_TT	Sn–Xv trail termination source and sink functions
Snm_TT	Sn non-intrusive monitor trail termination sink function
Sns_TT	Sn supervisory-unequipped termination source and sink functions
SnD_TT	Sn tandem connection (option-2) termination source and sink functions
SnD/Sn_A	Sn tandem connection (option-2) to VC–n layer adaptation source and sink functions
SnDm_TT	Sn tandem connection (option-2) non-intrusive monitor sink functions
SnT_TT	Sn tandem connection (option-1) termination source and sink functions
SnT/Sn_A	Sn tandem connection (option-1) to VC–n layer adaptation source and sink functions
SnTm_TT	Sn tandem connection (option-1) non-intrusive monitor sink function
SnT/DL_A	Sn tandem connection (option-1) to data link adaptation source and sink functions
S4–Xc > S4–Xv_I	Contiguous VC–4–Xc to virtual VC–4–Xv concatenation interworking function
S4–Xv > S4–Xc_I	Virtual VC–4–Xv to contiguous VC–4–Xc concatenation interworking function
Sn/<client>_A	VC–n layer to <client> layer adaptation source and sink functions, the <client> signal can be any of:
	- Sm, lower order SDH VC–2, VC–12, VC–11 layer - Pqx, PDH E31x, E32x, E4x layer - Avp, ATM virtual path layer - EthP, Ethernet path layer - CBR-t, Constant bit-rate transport layer - User channel, F2 and F3 layer
<server>/Sn_A	<server> layer to VC–n layer adaptation source and sink function, the <server> signal can be: - MSn, SDH multiplex section (n = 1, 4, 16, 64, 256)

Note: In this example point-to-multipoint connections have to be provisioned as separate connections to the same input (T)CP.

Note: In case a network element supports 1 + 1 protected matrix connections in its Sn_C function, this function may contain at any moment in time either all unprotected matrix connections, or all 1 + 1 protected matrix connections, or a mixture of unprotected and 1 + 1 protected matrix connections. The actual set of matrix connections, associated connection types and directions is an operational parameter controlled by network management.

Symbol

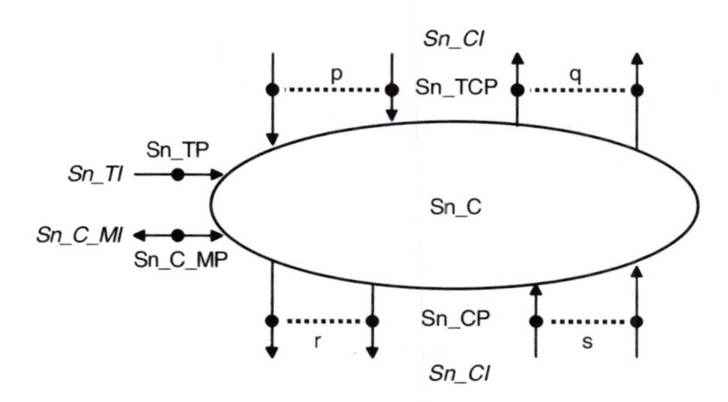

Figure 7.4 Uni-directional Sn_C symbol.

The number of available inputs and outputs is normally symmetrical, i.e. p = q and r = s.

Interfaces

Table 7.4 Sn_C input and output signal.

Inputs	Outputs
Per Sn_TCP and Sn_CP Sn_CI_Data Sn_CI_Clock Sn_CI_FrameStart	**Per Sn_TCP and Sn_CP** Sn_CI_Data Sn_CI_Clock Sn_CI_FrameStart
Once per function: Sn_TI_Clock Sn_TI_FrameStart	
Per input and output connection point: Sn_C_MI_ConnectionPortIds	
Per matrix connection: Sn_C_MI_ConnectionType Sn_C_MI_Directionality	
Per SNC protection group: Sn_CI_APS Sn_CI_SSF Sn_CI_SSD Sn_AI_TSF Sn_AI_TSD Sn_C_MI_PROTtype Sn_C_MI_OPERtype Sn_C_MI_WTRtime Sn_C_MI_HOtime Sn_C_MI_EXTCMD	**Per SNC protection group:** Sn_CI_APS

Processes

In the Sn_C function, VC–n layer characteristic information is routed between input (termination) connection points ((T)CPs) and output (T)CPs by means of matrix connections. (T)CPs may be allocated within a protection group.

Routing

The function shall be able to connect a specific input with a specific output by means of establishing a matrix connection between the specified input and output. It shall be able to remove an established matrix connection.

Each (matrix) connection in the Sn_C function should be characterized by the parameters shown in Table 7.5.

Table 7.5 Sn_Cconnection provisioning.

parameter	value
Sn_C_MI_ConnectionPortIds	Set of (T)CP identifiers, input, output
Sn_C_MI_ConnectionType	– unprotected – 1 + 1 protected – 1:1 protected – . . .
Sn_C_MI_Directionality	– uni-directional – bi-directional

Protection

If the connection function supports protection switching, additional parameters have to be provisioned as shown in Table 7.6.

Table 7.6 protection provisioning.

parameter	value
Sn_C_MI_PROTtype	Protection group selection, i.e. the set of connection points (T)CP, the protection architecture: 1 + 1/ 1:n/m:n, switching type: uni-/bi-directional, operation type: APS usage: true/false, extra traffic: true/false)
Sn_C_MI_OPERtype	revertive non-revertive
Sn_C_MI_WTRtime	0–12 minutes, in steps of 1 minute

Table 7.6 (*Continued*)

Sn_C_MI_HOtime	0–10 seconds, in steps of 100 ms
Sn_C_MI_EXTCMD	CLearRequest LockOut ForcedSwitch ManualSwitch EXERcise

The operation of the protection switch depends on the protection type and may be initiated either by signals from the layer itself, e.g. TSF and TSD, or from the server layer, e.g. SSF and SSD, by the APS protocol from the far-end, or upon request from the operator via the element management system.

Provided no protection switching action is activated/required, the following changes to (the configuration of) a connection shall be possible without disturbing the CI passing the connection:

- addition and removal of protection;
- addition and removal of connections to/from a broadcast connection;
- change between operation types;
- change of Wait-To-Restore time;
- change of Hold-off time.

Unequipped VC generation

The function shall generate an unequipped VC–n signal, as defined in ITU-T Rec. G.707 (2003). This is a signal having an all-zero value in the higher order virtual container path overhead signal label octet (C2), the tandem connection monitoring byte (N1) and the path trace byte (J1), and a valid BIP-8 byte (B3). The virtual container payload and the remaining path overhead bytes are unspecified.

Defects

None.

Consequent actions

If an output of this function is not connected to one of its inputs, the function shall connect the unequipped VC–n (with valid frame start (FS) and SSF = false) to the output. This may be implemented by providing a trail termination function UNEQ_TT_So that transmits the unequipped signal and that is connected to all unused outputs.

Defect correlations

None.

Performance monitoring

None.

7.2.2 *ETH flow domain*

The second example is the Ethernet flow domain defined in ITU-T Rec. G.8010 (2004) clause 6.3.2.2.

The Ethernet flow domain ETH_FD is defined as the functionality that is available to transfer Ethernet characteristic information ETH_CI between a set of (termination) flow points ETH_FP and/or ETH_TFP. The association between one ingress and one or more egress ETH (termination) flow points, provisioned to transfer ETH_CI traffic units across the ETH flow domain, need not be present at all times. In general, ETH flow domains may be partitioned into smaller flow domains interconnected by ETH flow point pool links as described in Chapter 3, Section 3.2.4. The matrix (e.g. bridge) is a special case of an ETH flow domain.

An ETH flow domain provides broadcast connectivity between the connected ETH (termination) flow points. An ETH_CI traffic unit received via an input port (e.g. A in Figure 7.5) of the ETH flow domain is forwarded to all output ports on the ETH flow domain (B, C, D), with

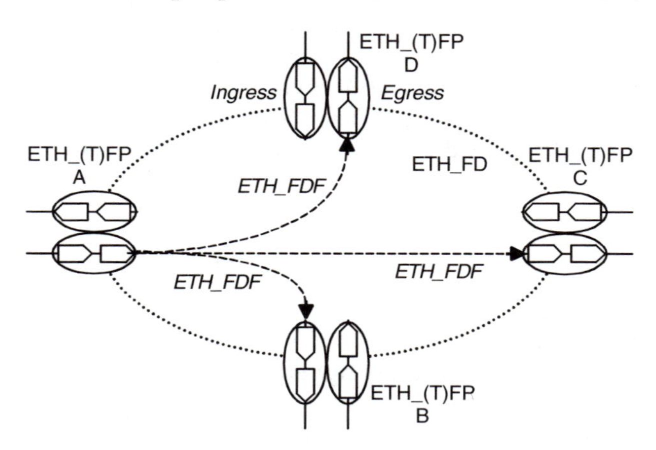

Figure 7.5 ETH flow domain connectivity.

the exception of the output port (A) that is in the same bi-directional ETH (termination) flow point as the input port.

By means of ETH network management, ETH control plane actions and/or MAC learning, connectivity for a particular ETH flow point can be restricted.

7.3 CONNECTION MATRIX EXAMPLES

Generally, the connection function described in the previous sections is highly flexible and will provide full flexibility between its input ports and output ports as illustrated in Section 7.3.1. In general, the connection matrix has n input ports and n output ports, i.e. it is an $n \times n$ matrix. However, due to implementation constraints the connectivity may be affected. Some of the limitations could be:

- Only bi-directional connections are supported: when input[i] is connected to output[j] at the same time input [j] will be connected to output[i].
- Only point-to-point connections are supported: when input[i] is connected to output[j] it is not possible to connect input[i] also to output[k].
- Blocking in a multistage connection matrix: depending on the number of connections already present in the matrix it is not always possible to connect input[i] with output[j].
- No connections within a group of ports: e.g. input[i] cannot be connected to output[j] if they are both tributary ports.
- No rearrangement of the multiplex structure: if the server layer uses multiplexing for the transport of multiple client signals the order of the client signal in the multiplex (e.g., time slots, frequency slots, wavelength slots) cannot be interchanged, i.e. input[i] can only be connected to output[i].

It is possible to represent the limited connectivity by grouping input and output ports and describe the connectivity between the ports as illustrated in the following examples.

7.3.1 Connection matrix example for full connectivity

This is the most generic example of a matrix. In this matrix there are no restrictions to the connectivity. Every input port can be connected to

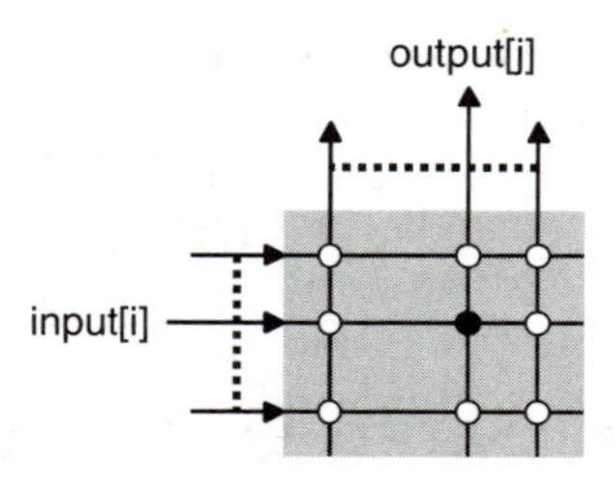

Figure 7.6 Matrix with full connectivity.

every output port as indicated by the white dots, a black dot indicates that a connection is provisioned between input[i] and output[j]. It would even be possible to connect an input to more than one output. Figure 7.6 shows the conventional uni-directional way of drawing a matrix.

The actual implementation of this matrix can be a multistage non-blocking Clos matrix to limit the number of required connection points. Table 7.7 is an alternative way to show the full connectivity of the matrix.

Table 7.7 Matrix with full connectivity.

	output[j]
input[i]	×

× indicates that a matrix connection is allowed between every input[i] and output[j].

Figure 7.7 represents the functional model of this matrix. Note that no distinction is made between individual inputs or between individual outputs.

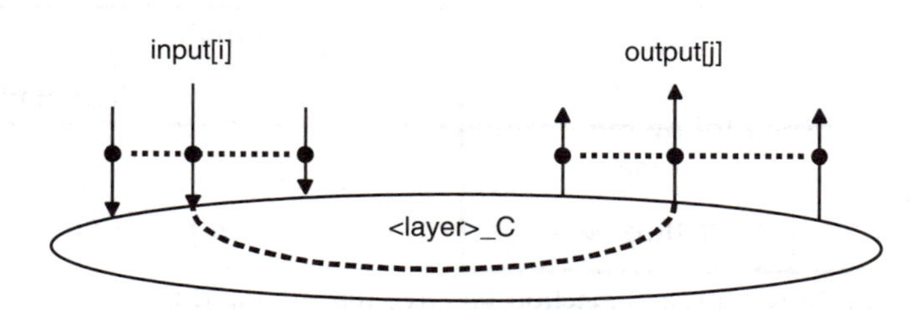

Figure 7.7 Functional model of matrix with full connectivity.

7.3.2 *Connection matrix example for two groups*

In this example the total set of inputs and outputs is divided into two distinctive groups: a line group and a tributary group. The line group contains the inputs and outputs of the signals that are transported over the network and acts as the server of the client signals passing the inputs and outputs in the tributary group. Normally, this matrix only supports the provisioning of connections between the line group and the tributary group and does not support connections restricted to only the line group or tributary group. However, some implementations allow loopbacks, i.e. connections provisioned between the input and output of the same line or tributary signal. Loopbacks can be transparent where the input[i] remains connected to the original output[j] as provisioned before the loopback was applied, or non-transparent where the signal at output[j] is replaced by an 'idle', e.g. AIS signal. Figure 7.8 illustrates this matrix with loopback capability.

Table 7.8 represents the connectivity of this matrix.

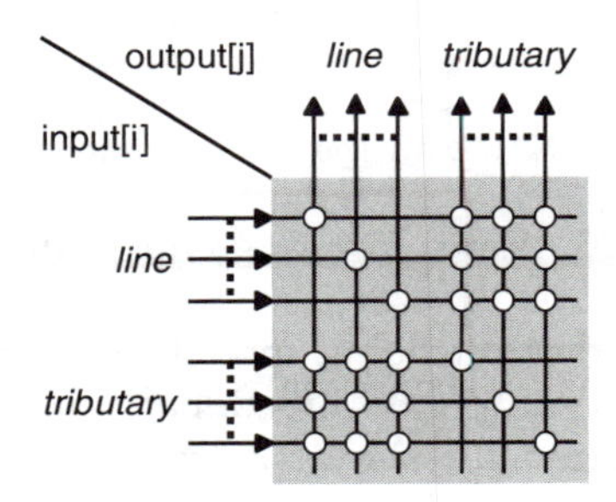

Figure 7.8 Matrix for two groups.

Table 7.8 Connection matrix for 2 groups.

		Output[j]	
		Line	**Tributary**
Input[i]	**Line**	i = j	×
	Tributary	×	i = j

× Indicates matrix connection possible for any i and j.
i = j Indicates matrix connection only possible for i = j (i.e. loopback).

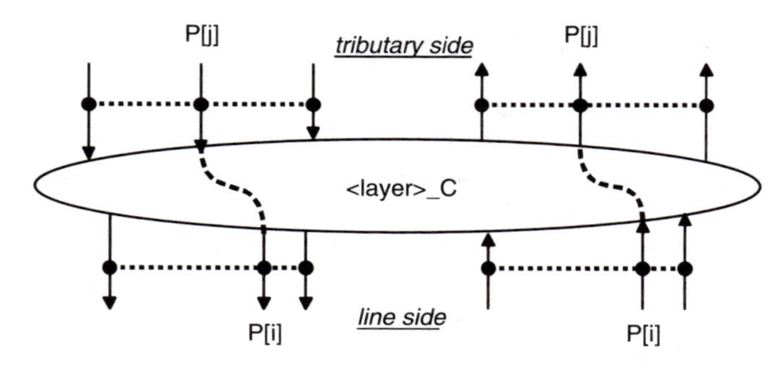

Figure 7.9 Matrix for two groups with bi-directional connection.

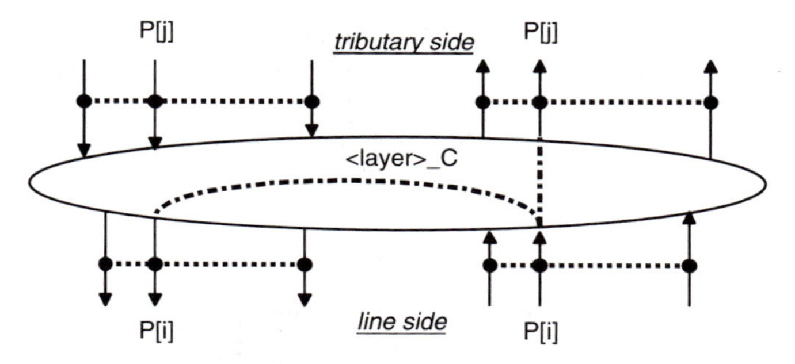

Figure 7.10 Matrix for two groups with loopback.

Figure 7.9 shows the functional model of a connection matrix for two groups with an example bi-directional connection from line[i] to tributary[j] and Figure 7.10 shows a transparent line loopback for line[i] connected to tributary[j].

7.3.3 *Connection matrix example for three groups*

In this example the total set of inputs and outputs is divided into three groups, each group containing both inputs and outputs. Generally, the matrix provided here is present in network elements that are part of a network with a ring structure. To make a distinction between the lines that connect to the two adjacent network elements they are identified as 'line west' and 'line east'. Since not all signals received at the line side are terminated locally but are passed through the tributary, signals are generally referred to as add/drop signals. The network element is referred to as an *Add/Drop Multiplexer* ADM. An example of a ring topology with four ADMs is depicted in Figure 7.11.

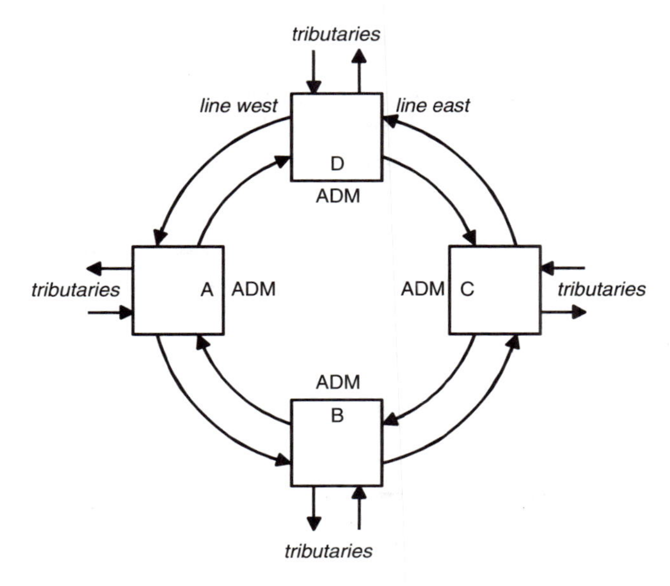

Figure 7.11 Example ring topology.

Generally, it is not useful to provide connection capabilities within the line groups for normal transport, as this would create a loopback. Also, tributary signals are generally transported to other network elements and do not require connectivity within the group. Instead of using a matrix with full connectivity as described in Section 7.3.1, this application requires only a matrix with restricted capabilities. This is illustrated in Figure 7.12.

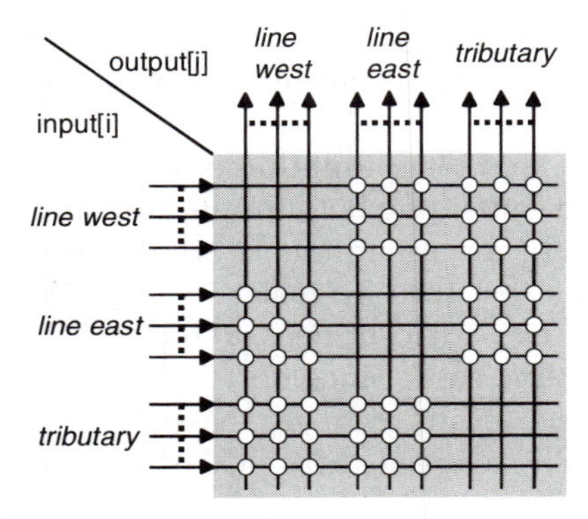

Figure 7.12 Matrix for three groups.

If protection of the lines is required the connection capability of the matrix has to be extended based on the protection methodology. For example, Chapter 9, Section 9.2.1.2 describes the multiplex section protection process and its requirements for connectivity.

Table 7.9 provides the alternative method to show the connectivity.

Table 7.9 Connection matrix for three groups.

		Output[j]		
		line west	**line east**	**tributary**
Input[i]	**line west**	–	×	×
	line east	×	–	×
	tributary	×	×	–

× Indicates matrix connection possible for any i and j.
– Indicates no connection possible.

Line side connectivity restrictions

The connectivity of this matrix can even be further restricted by allowing only connections between line east and line west for inputs and outputs with the same index, e.g. input[2] and output[2]. This changes the matrix in Figure 7.12 into the one depicted in Figure 7.13 and Table 7.10.

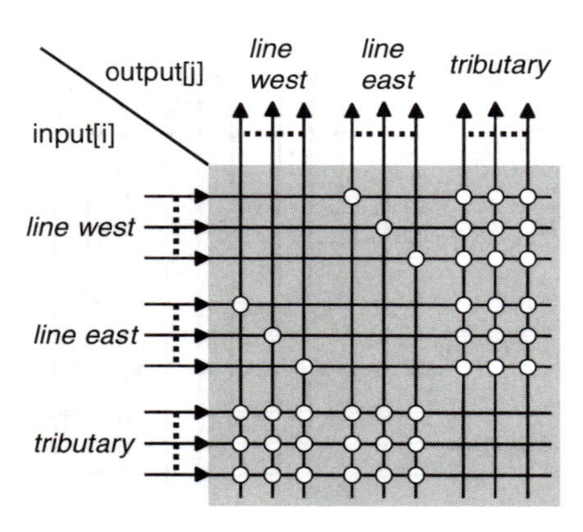

Figure 7.13 Matrix for three groups, restricted.

Table 7.10 Connection matrix for three groups; line access restrictions.

		Output[j]		
		line west	**line east**	**tributary**
Input[i]	**line west**	–	i = j	×
	line east	i = j	–	×
	tributary	×	×	–

× Indicates matrix connection possible for any i and j.
– Indicates no connection possible.

Tributary side connectivity restrictions

Another form of restriction is also possible. Instead of restricting the connectivity at the line side of the matrix, the add/drop capability is restricted to a limited number of line inputs and outputs. In this example, the tributaries can be connected only to input[1] and output[1] of the east and west line. This changes the matrix in Figure 7.12 into the one depicted in Figure 7.14 and Table 7.11.

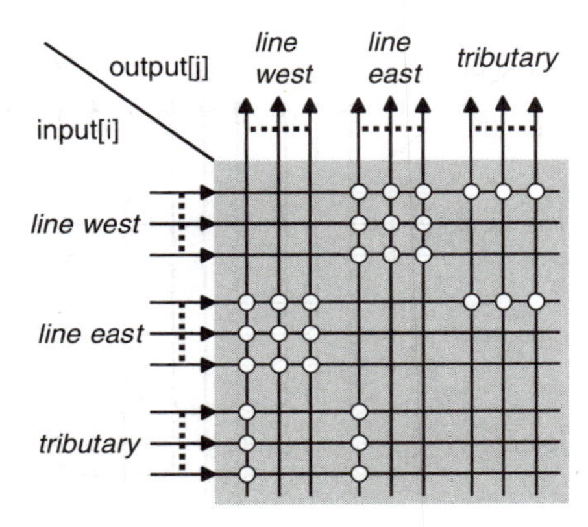

Figure 7.14 Matrix for three groups, trib. access restricted.

Table 7.11 Connection matrix for three groups; trib. access restriction.

		Output[j]		
		line west	**line east**	**tributary**
Input[i]	**line west**	–	×	i = 1
	line east	×	–	i = 1
	tributary	j = 1	j = 1	–

× Indicates matrix connection possible for any i and j.
– Indicates no connection possible.

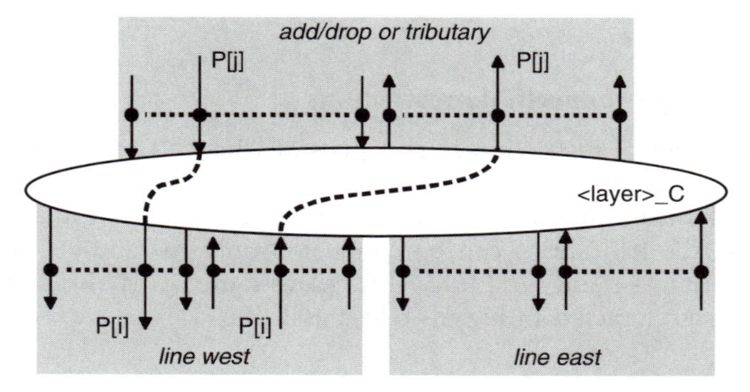

Figure 7.15 Functional model of matrix for three groups.

The functional model of a connection matrix for three groups is provided in Figure 7.15. Note that the restrictions are not indicated in this model.

8

Connection supervision

Performance monitoring

One of the criteria that was set right at the start of the development of the SDH and SONET multiplexing methodology was that the 'Quality of Service' (QoS) should be improved. The PDH multiplexes provided hardly any possibility of checking the QoS for the transported signals. The addition of the 'Cyclic Redundancy Checking' (CRC) to the 2.048 kbit/s signals defined in ITU-T Rec. G.706 (1991) were a step towards providing QoS in PDH. To monitor the integrity of the transported signals in SDH, SONET, OTN, etc., specific methodologies can be applied. This chapter will describe some of the methods that can be used to get information about the QoS or performance of a trail or connection.

8.1 QUALITY OF SERVICE

Supervisory processes are required to assess the Quality of Service. These processes analyze the actual occurrence of a disturbance or fault of the transported signal with the purpose of providing an appropriate indication of the performance of the information transfer to the network operator. This type of performance should not be confused with the performance regarding clock accuracy, jitter and wander, signal amplitude, and eye-patterns, that is, properties that are normally monitored by external measuring equipment. Generally, the process of monitoring the error performance consists of the collection, analysis,

SDH/SONET Explained in Functional Models Huub van Helvoort

and reporting of performance information. Normally, each transport entity has its own associated monitoring process. All the performance processes should run continuously to provide the most reliable and up to date information. The performance information will be retrieved from trail termination, adaptation and connection functions in the form of performance parameters. The parameters are accumulated during predetermined periods of time. Which functions are capable of gathering, storing, thresholding and reporting these parameters is technology dependent. The performance history can be correlated and used to determine whether part of a network element has to be replaced because it has degraded to a preset level. It can also be interpreted and used to verify that *Service Level Agreements* (SLA) between network operators and clients are met. The latter application of performance monitoring is related mainly to network connections; this will be elaborated in the following sections.

8.2 CONNECTION MONITORING METHODS

In this section, four methodologies are described that can be used to monitor the performance of a connection or a trail in a network, i.e. by inherent monitoring, non-intrusive monitoring, intrusive monitoring or sublayer monitoring.

Figure 8.1 shows only a specific fragment of a transport network, i.e. the part that is sufficient to illustrate the methodologies explained in this chapter. It is the part that is located between a client trail and a

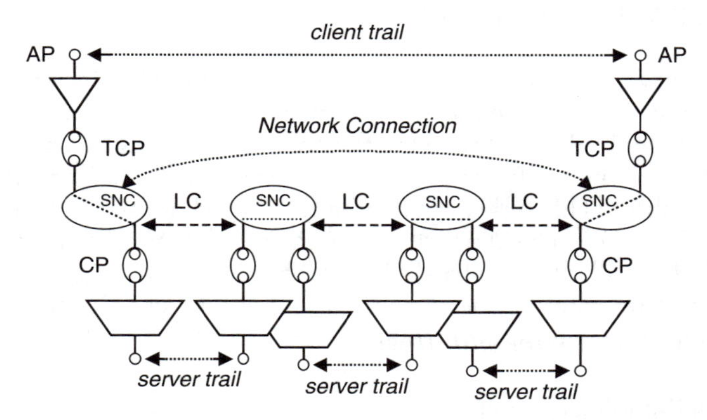

Figure 8.1 Functional model fragment.

server trail. Figure 8.1 shows four network elements, two elements are the nodes where the client trail starts and ends and two elements are possible intermediate nodes. The intermediate nodes are included to illustrate the possibility that there is more than one server trail.

8.2.1 *Inherent monitoring*

This methodology monitors trails indirectly. In this case, the performance data that are available inherently from the server layer network are used. Figure 8.1 can be used to illustrate this method because no additional functionality is required.

If one of the associated server trails in one of the server layer networks fails it may provide an indication of the failure (e.g. AIS) at the output of the link connection LC that is supported by the server trail. This indication is forwarded, in sequence, over the next link connection that is supported by other trails in the server layer. The output of the last link connection in the network connection may provide a trail signal fail indication.

Note that in SDH and OTN, the output of each Sn or ODUk link connection is able to detect the AIS to indicate that the transport through one of the preceding trails in the server layer has failed. The outputs of link connections in ATM and PDH are not able to detect this failure indication.

It is possible to collect error performance information for each individual link connection in the server layer network. When the server layer adaptation function multiplexes more than one client signal, the error performance information for each of the link connections, supported by the server layer trail, will not be available individually; it must be estimated from the error performance of the server trail. Processes in the management network may correlate the performance information collected from each link connection that is part of the overall connection of interest. Since it does not include the performance of the server layer adaptation and connection functions the overall error performance or QoS of the client trail cannot be provided by this monitoring methodology.

8.2.2 *Non-intrusive monitoring*

This methodology monitors a client trail directly. In this case special termination functions monitor the characteristic information of the

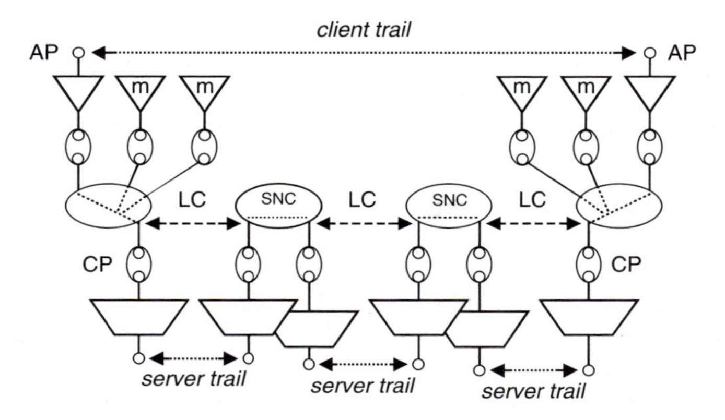

Figure 8.2 Non-intrusive monitoring.

client trail. These special termination functions only have an input for the signal to be monitored non-intrusively and are generally referred to as monitoring termination functions <*signal*>m_TT. They are connected to the client signal in the connection function as illustrated in Figures 8.2 and 8.3. The latter is a uni-directional representation of the connection function depicting the actual matrix connections required to connect the non-intrusive monitors.

The performance information collected by the monitoring trail terminations can be retrieved to provide the QoS of the client trail.

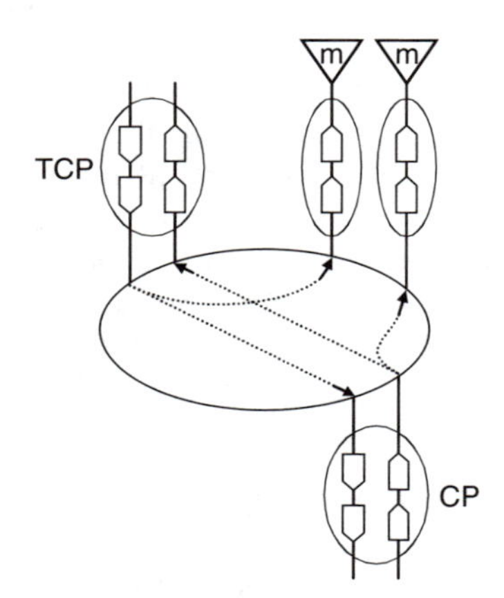

Figure 8.3 Connection configuration.

The retrieved information concerns the connection from the originating trail termination source function up to the point where the monitor is connected. It is also possible to monitor a specific part of a network connection by connecting monitors in connection functions of intermediate nodes. Processes in the management network may correlate the performance data retrieved from these intermediate monitors to derive the performance of a specific section, e.g. a section that is part of a different operator's network. It is even possible to use this correlation methodology to monitor arbitrary nesting or overlapping of network connection sections. Apart from the performance, the connectivity of a segment can also be checked by the monitors and used for fault localization. However, this requires proper identification of the monitored signal; in SDH the *Trail Trace Identifier* (TTI) is used for this purpose.

8.2.3 *Intrusive monitoring*

This methodology monitors a normal, payload carrying client trail by interrupting it to replace it by a test signal originating from a test trail termination function <*signal*>t_TT function. This is depicted in Figures 8.4 and 8.5. The test trail can replace the client trail at any arbitrary section of the network connection. Note that in order to switch simultaneously this requires the synchronization of the connection functions at both source and sink side of the test trail.

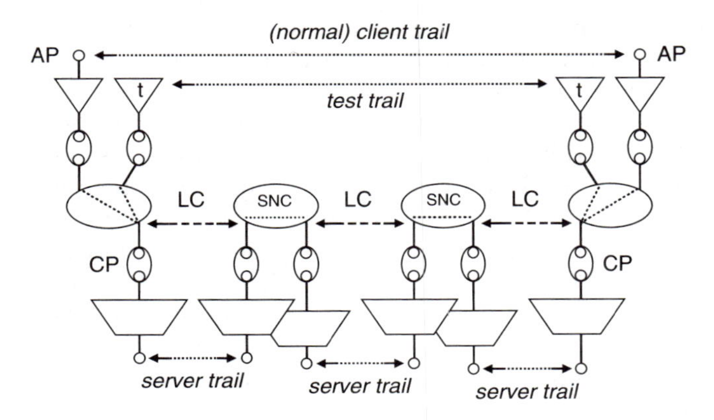

Figure 8.4 Intrusive monitoring.

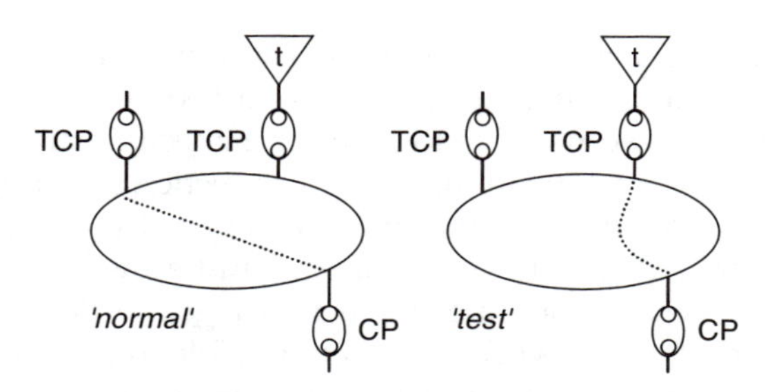

Figure 8.5 Normal and test connection.

Even though this method allows all parameters to be monitored directly, the disadvantage is that the client trail is interrupted and thus affects the availability of client signal, i.e. the QoS is degraded. Intrusive monitoring can only be performed during the setup phase of a client trail, e.g. to test the performance or connectivity of the path through the network.

This method also supports arbitrary nesting or overlapping of the network connections, but no simultaneous testing.

8.2.4 Sublayer monitoring

This methodology uses the functional modeling capability to create a sublayer by expanding a connection point as described in Chapter 4, Section 4.2.3. Since a connection point is associated with a connection function, the sublayer adaptation function <sublayer>/<client>s_A and sublayer trail termination function <sublayer>s_TT can be depicted either collapsed (as shown in Figure 8.6) or expanded (as shown in Figure 8.7). The <sublayer>s_TT function has the capability to monitor the sublayer trail. This capability depends on the used technology, e.g. in SDH it uses a dedicated part of the trail OA&M overhead and in packet networks it may use specific OA&M packets. The sublayer can be created in any two nodes in the network connection to monitor the enclosed section. Intermediate nodes located between these two nodes shall transfer the specific OA&M transparently.

This method allows all performance parameters to be tested directly assuming that the used technology offers this OA&M capability. This method also allows the nesting of monitored sublayer trails again

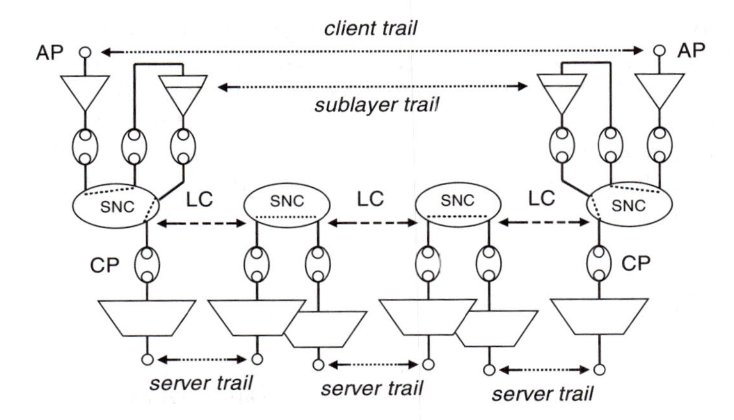

Figure 8.6 Sublayer monitoring.

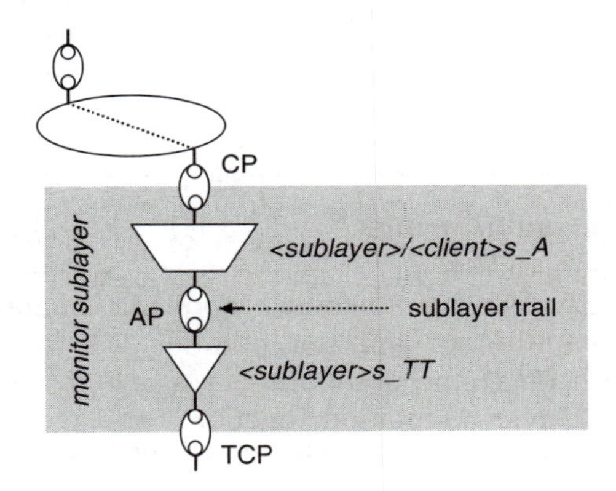

Figure 8.7 Monitoring sublayer.

assuming that the technology supports this nesting, e.g. SDH allows one level of nesting.

8.3 CONNECTION MONITORING APPLICATIONS

As already mentioned in the introduction of this chapter, performance monitoring is used to check and guarantee the desired QoS, especially for existing connections. However, to avoid connecting a client signal to a network connection that is already degrading, unused connections can be monitored for their quality. When a trail does not run completely in a single operator's network but passes another operator's network, service level agreements (SLA) between the two are closed. Part of such an agreement will be the delivered QoS and this has to be monitored from ingress to egress of part of the trail, i.e. the tandem connection.

8.3.1 Monitoring of unused connections

A connection is unused if at least one of the access points that delimits the trail connection is not bound to a client signal adaptation function (see Figure 8.8).

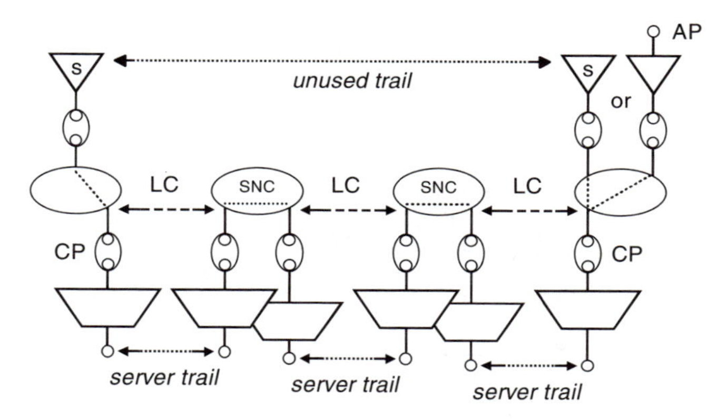

Figure 8.8 Monitoring unused connection.

A supervisory unequipped trail termination function <*trail*>s_TT can be used to monitor unused connections. The source function transmits a layer characteristic information structure with an undefined payload and valid trail overhead; the sink function reports the performance monitoring information to the element management system. Remote information is exchanged between the sink and source function as shown in detail in Figure 8.9.

Note that any regular trail termination function may be used as a supervisory trail termination function if it can be operated without a payload signal.

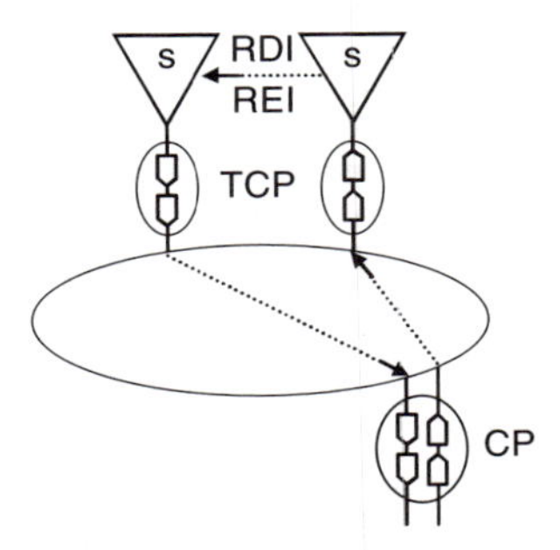

Figure 8.9 Supervisory unequipped.

8.3.2 Tandem connection monitoring

A tandem connection monitoring provides the capability to monitor a section of a trail independent of the performance monitoring of the complete trail, as shown in Figure 8.10.

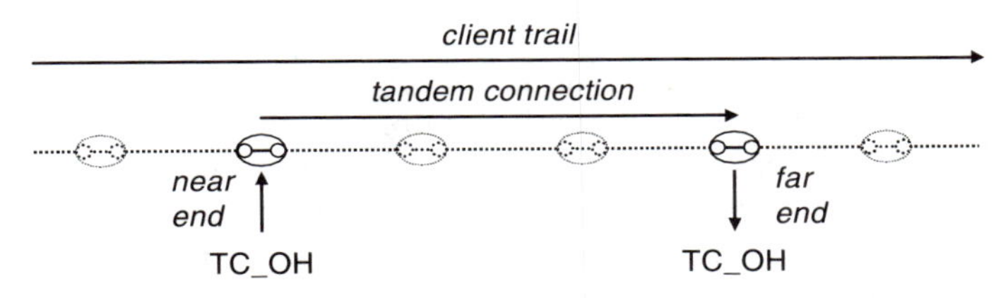

Figure 8.10 Tandem connection monitoring.

To be able to offer this capability, the tandem connection process has to meet the following requirements:

- The tandem connection specific overhead (TC-OH) shall be inserted at the near-end and removed at the far-end of a tandem connection.
- The tandem connection process at the near-end shall report the performance and failure/alarm conditions related to the incoming signal for the purpose of fault localization; this information is also forwarded to the far-end in the TC-OH.
- The tandem connection process at the far-end shall report the performance and failure/alarm conditions related to the incoming signal for the purpose of fault localization, e.g. by correlating the bit-error rate detected at near-end and far-end, the bit-error rate of the tandem connection can be derived.

- The tandem connection monitoring shall be independent of any incoming server signal fail indication, e.g. AIS, FDI.
- The tandem connection process at the near-end shall report and forward a detected incoming server signal fail indication for the purpose of fault localization, e.g. before or after the near-end.
- The tandem connection process at the far-end shall report and forward a detected incoming server signal fail indication for the purpose of fault localization, e.g. before or after the far-end.
- The tandem connection connectivity between near-end and far-end shall be traceable, i.e. each tandem connection shall have its own specific *Access Point Identifier* (API).
- The tandem connection process shall report continuity defects, i.e. loss of signal, unequipped signal, and API mismatch, detected between the near-end and the far-end.

Several applications of tandem connections are possible; three possible applications are described below:

- A tandem connection can monitor the quality of the service delivered to a client. This is an application in a public network domain, a network operator domain, a network operator sub-network domain, etc. Generally it is referred to as a *serving operator administrative domain*. The functional model is depicted in Figure 8.11. In this application the supporting tandem connection has its sublayer source functions as close as possible behind the *user to network interface* (UNI) or *network to network interface* (NNI), and its sublayer sink functions as close as possible in front of the UNI or NNI.

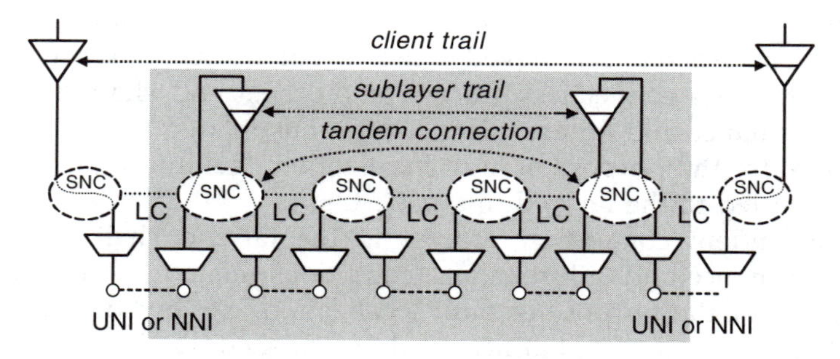

Figure 8.11 Serving operator administrative domain.

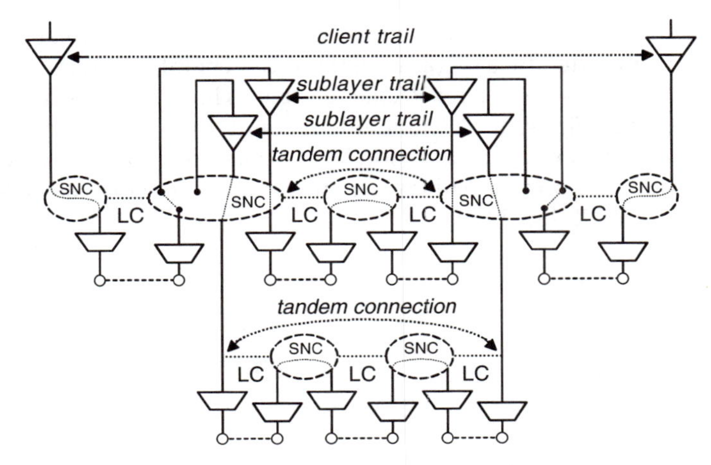

Figure 8.12 Protective domain.

- Two tandem connections can be used to monitor the defect status of the working and protection network connections. Normally, this application is called a *protected domain*; the sublayer monitored SNC protection methodology is described in Chapter 9, Section 9.2.2.1. The functional model of this application is shown in Figure 8.12. A protected domain utilizing tandem connections has its sublayer sources right behind the protection switch bridge and its sublayer sinks just in front of the protection switch selector functions.
- A tandem connection can monitor the quality of the service of a section of a client trail. In a user domain this can be the service

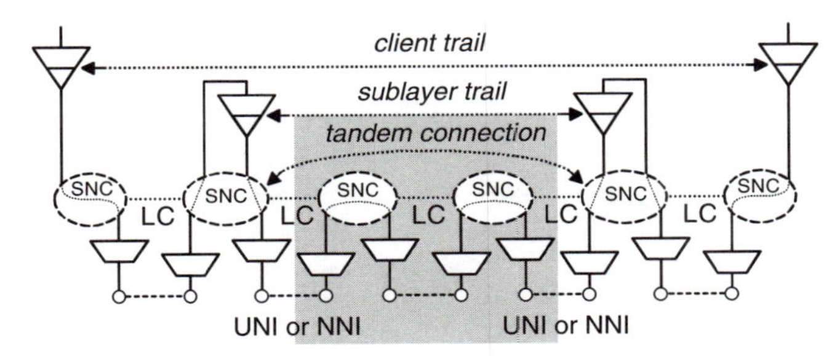

Figure 8.13 Service requesting administrative domain.

provided by the operator; in an operator domain this can be the service provided by a second operator. In general, it is referred to as a *service requesting administrative domain*. The functional model is provided in Figure 8.13. The tandem connection in this application has its sublayer source function as close as possible in front of the UNI or NNI, and its sublayer sink function as close as possible behind the UNI or NNI.

9

Protection models

This chapter describes the functional models that can be used to model the protection schemes in a layered network. Generally, these protection schemes are used to improve the availability of a client signal. The availability is one of the components of a service agreement between operators or a performance requirement for network equipment.

9.1. INTRODUCTION

The modeling methodology used to describe a transport network can be utilized to illustrate one of the main strategies that are used to improve the availability of such a network. Availability is the ability of a transport network to be in such a state that it can transfer information under required performance conditions during a given time interval, assuming that the required resources are provided or, at its simplest level:

$$\text{Availability} = \text{Uptime}/(\text{Uptime} + \text{Downtime})$$

The availability of a telecommunications network or a trail in a transport network is ideally 100%, however, due to physical properties and human errors the availability is usually lower if no special measures are taken. From a design or implementation point of view the availability can be expressed as:

$$\text{Availability} = \text{MTBF}/(\text{MTBF} + \text{MTTR})$$

SDH/SONET Explained in Functional Models Huub van Helvoort

where MTBF is the *Mean Time Between Failures* of the used transport entities and MTTR is the *Mean Time To Repair* those resources. The availability can be improved by providing a replacement for the failed or degraded transport entities, such as equipment or interconnecting fibers. Now, the availability is determined mainly by the time it takes to replace the failed or degraded resources, i.e. the MTTR. Decreasing the MTTR will increase the availability of the network. The use of these replacing transport entities is normally a consequent action based on detection of a defect or of a degraded performance. The replacement can also be based on an external request, e.g. a command issued by an operator at a network management system.

Two types of replacement can be distinguished:

- **Replacement by protection.**
  This type of replacement utilizes pre-allocated capacity between two or more nodes in the network that is dedicated to protection. Normally, a stand-by or *protection transport entity* will replace the original, or *working transport entity*. The architecture dedicated to this type of replacement can have different grades of complexity. In its most simple form the architecture has a single dedicated protection entity to replace each working entity $(1 + 1)$ and is normally referred to as *one-to-one protection* (see Figure 9.1).

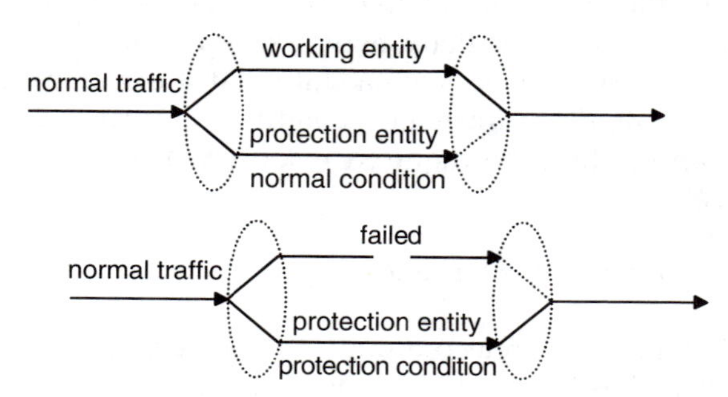

Figure 9.1 $1 + 1$ protection.

Normally, this form of protection is single ended because the selection between working or protection entity is performed at the receiving end only. At the sending end the information is distributed or bridged over both working and protection entities.

When the protection entity is not in use because the working entity is carrying the client information or normal traffic, its

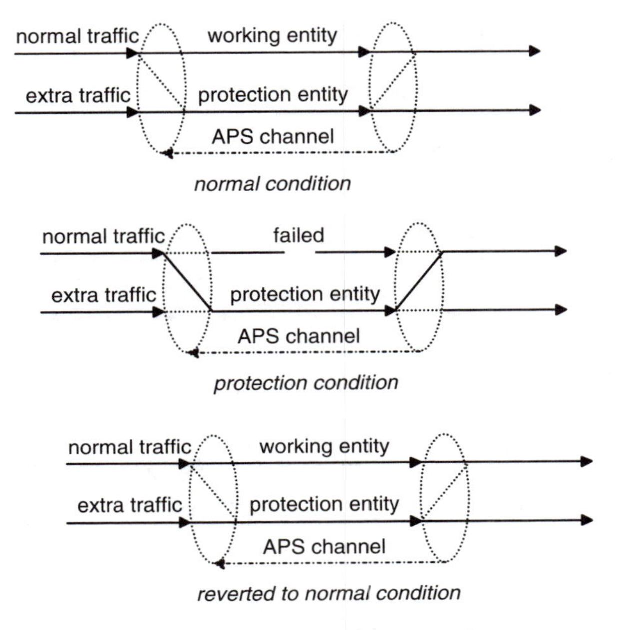

Figure 9.2 1:1 protection; revertive.

capacity can be used to transport extra traffic. This type of protection is referred to as (1:1) or *one-by-one protection* and is illustrated in Figure 9.2.

Now, the sending end has to know which transport entity the receiving end has selected. For this purpose an *Automatic Protection Switch* (APS) channel has been defined and an APS protocol was developed to align the receiving and sending end of the transport entity.

When the failed working entity has been repaired this will be detected at the receiving end and normally the protection switch returns to its original position: the protection switch is revertive as illustrated in Figure 9.2. However, this will most probably cause a hit, i.e. bit-errors, in the transported information.

In case this is not acceptable, it is possible to implement or provision the protection switch to be non-revertive: after each protection switch the designation of working and protection to the interconnecting entities will be switched as well. Figure 9.3 shows the non-revertive protection switch.

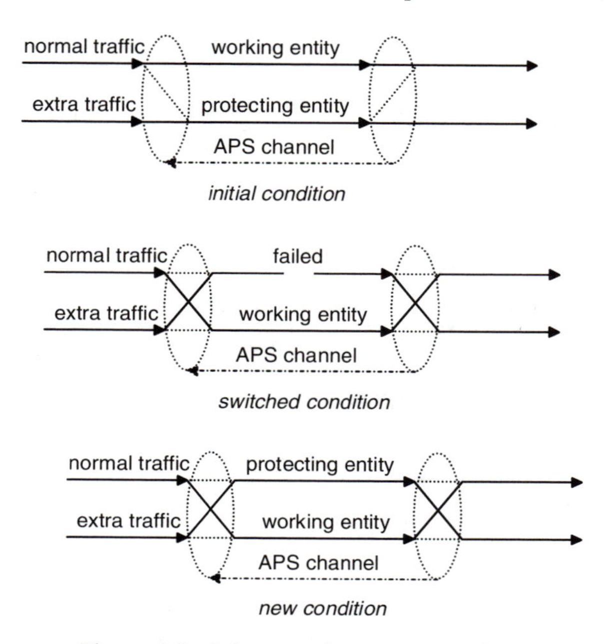

Figure 9.3 1:1 protection; non-revertive.

In its most complicated form the architecture has m separate protection entities to replace n working entities (m:n), normally referred to as *n by m protection* (see Figure 9.4).

Replacement or protection switching can be either uni-directional or bi-directional. Uni-directional protection switching replaces the affected transport entities only in the affected transport direction in the case of a uni-directional failure. Bi-directional protection switching replaces not only the affected transport entity in the failed directions but also replaces the (unaffected) transport entity in the opposite direction.

- **Replacement by restoration.**
  This type of replacement utilizes any spare capacity available between nodes in the network. In general, restoration requires algorithms that utilize re-routing of transport entities. When restoration is used in a transport network a certain percentage of the network capacity has to be reserved for the re-routing of the working traffic. Since this type of replacement cannot be described using the functional model it will not be discussed in more detail in this book.

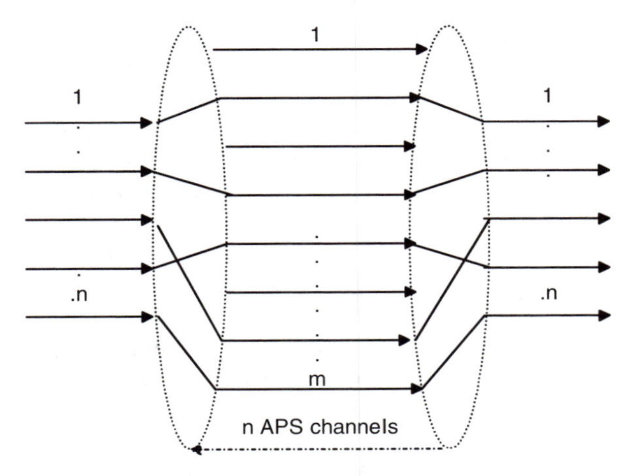

Figure 9.4 m:n protection.

9.2. PROTECTION

When SDH technology was introduced new network topologies were also introduced in the transport domain, one of the most remarkable being the ring. An SDH ring is comprised of a number of network elements that allow traffic to ingress, to egress the ring or pass through the ring. This equipment is commonly referred to as *Add/Drop Multiplexers* (ADM). The ADMs are interconnected in a loop configuration as illustrated in Figure 9.5.

This network topology is particularly suitable for providing protection to improve the availability of network resources.

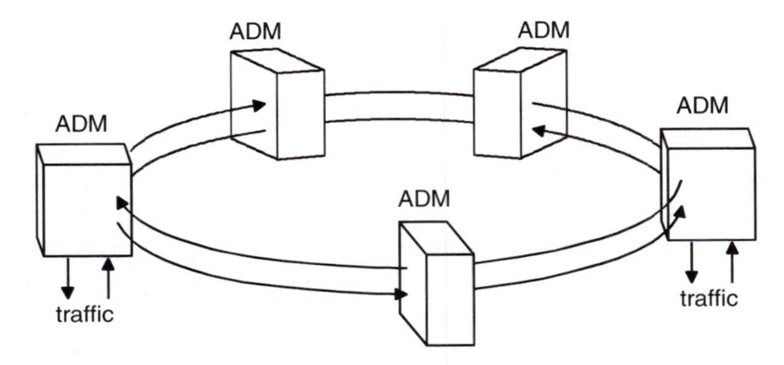

Figure 9.5 Typical SDH ring architecture.

Within the transport network architecture two distinctive types of protection can be identified: *trail protection* and *sub-network protection.*

Protection architectures and mechanisms are defined and described in separate recommendations, for SDH in ITU-T Rec. G.808.1 (2003) and for OTN in ITU-T Rec. G.873.1 (2003). ITU-T Rec.G.842 (1997) describes the interworking of SDH protection architectures.

9.2.1 *Trail protection*

Trail protection is a protection method applied in a transport layer network when a defect condition is detected in the same layer network (i.e. protection switching is activated in the same transport layer network).

A signal selected to be protected, i.e. the *protected trail*, consists of the (original) *working trail* that will be replaced by the *protecting trail* if the working trail fails or if its performance falls below the required level. The required level is usually described in an SLA (service level agreement) between a server (operator) and a client (customer). Two types of trail protection will be described:

1. path level, protecting a network connection; and
2. multiplex section or line level, protecting the sections in a ring of nodes in a network.

9.2.1.1 Path level trail protection

This type of trail protection is modeled by introducing a protection sublayer. The original trail termination (TT), as shown in the shaded area in Figure 9.6, is expanded according to the rules given in Chapter 4, Section 4.2.2.

The result is shown in Figures 9.7a and 9.7b for $1+1$ protection and Figures 9.8a and 9.8b for 1:1 protection. The protection adaptation functions A_p, the unprotected trail termination functions TT_u, the protected trail termination functions TT_p and a protection connection function C_p are introduced to model the switching between the protecting and working connections. The introduced adaptation, trail termination and connection functions are bound by the additional protection access points AP_p, the protection termination connection points TCP_p and the connection points CP. The status of the trails in the protection sublayer is made available to the protection connection

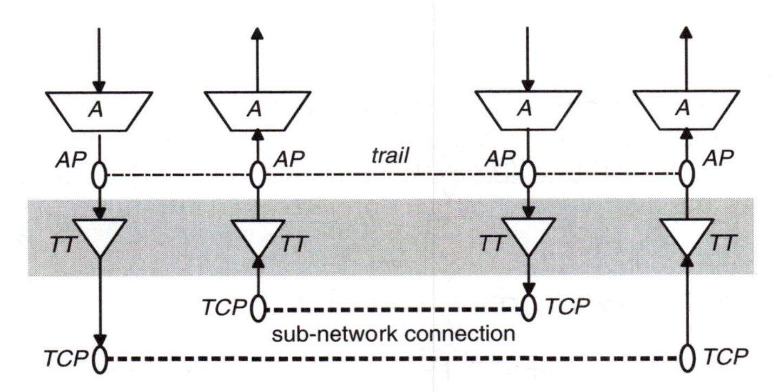

Figure 9.6 Unprotected trail.

function C_p, i.e. the Trail *Signal Fail/Degrade* (SF/SD) signal in Figures 9.7a, 9.7b, 9.8a and 9.8b by the unprotected trail termination TT_u. If the protection mechanism requires communication between the control functions of the protection connection functions, the protection adaptation function may provide access to an Automatic Protection Switch (APS) channel. The protected trail termination TT_p provides the status of the protected trail.

Figures 9.7a and 9.7b show the model for 1 + 1 trail protection. The shaded area indicates the protection sublayer that replaces the trail termination functions shown in the shaded area in Figure 9.6. As mentioned before, this type of protection is normally single-ended or uni-directional. The receiving end determines whether the working or protecting (sub-)network connection is used and if the use of the APS channel is not required. If both forward and return trail switching is required simultaneously at the near-end and far-end, i.e. bi-directional protection switching, the APS channel has to be used. In this case the protection switching protocol can be optimized depending on the predominant protection switching methodology (1 + 1 or 1:1) (for more details, see ITU-T Rec. G.841 (1998)).

Figures 9.8a and 9.8b show the model for 1:1 trail protection. The shaded area indicates the protection sublayer that replaces the trail termination functions shown in the shaded area in Figure 9.6 for both the normal traffic trail and for the extra traffic trail.

Normally, this type of protection is bi-directional and uses the APS channel for synchronizing the near-end and far-end of the trails. In the case of uni-directional use, the protection switch is based on local conditions and requests only; the near-end and far-end operate

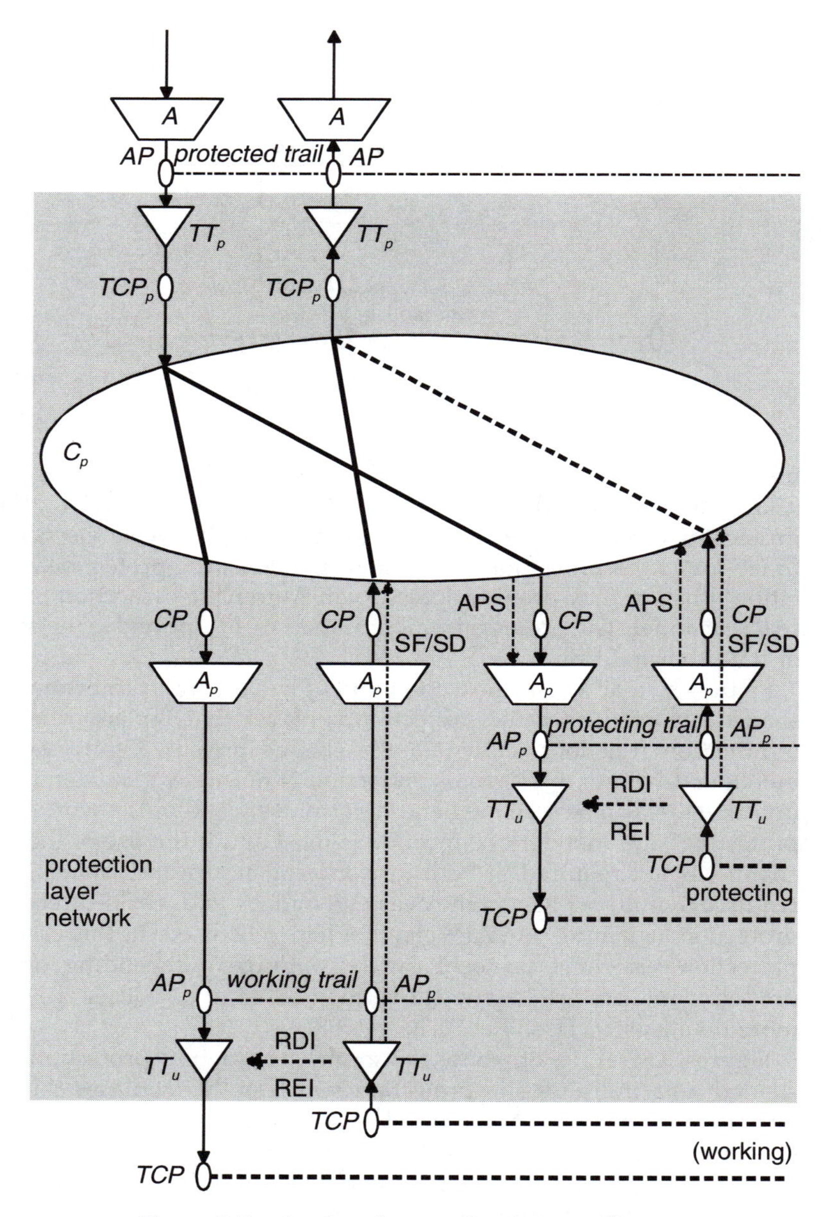

Figure 9.7a 1 + 1 trail protection (near-end).

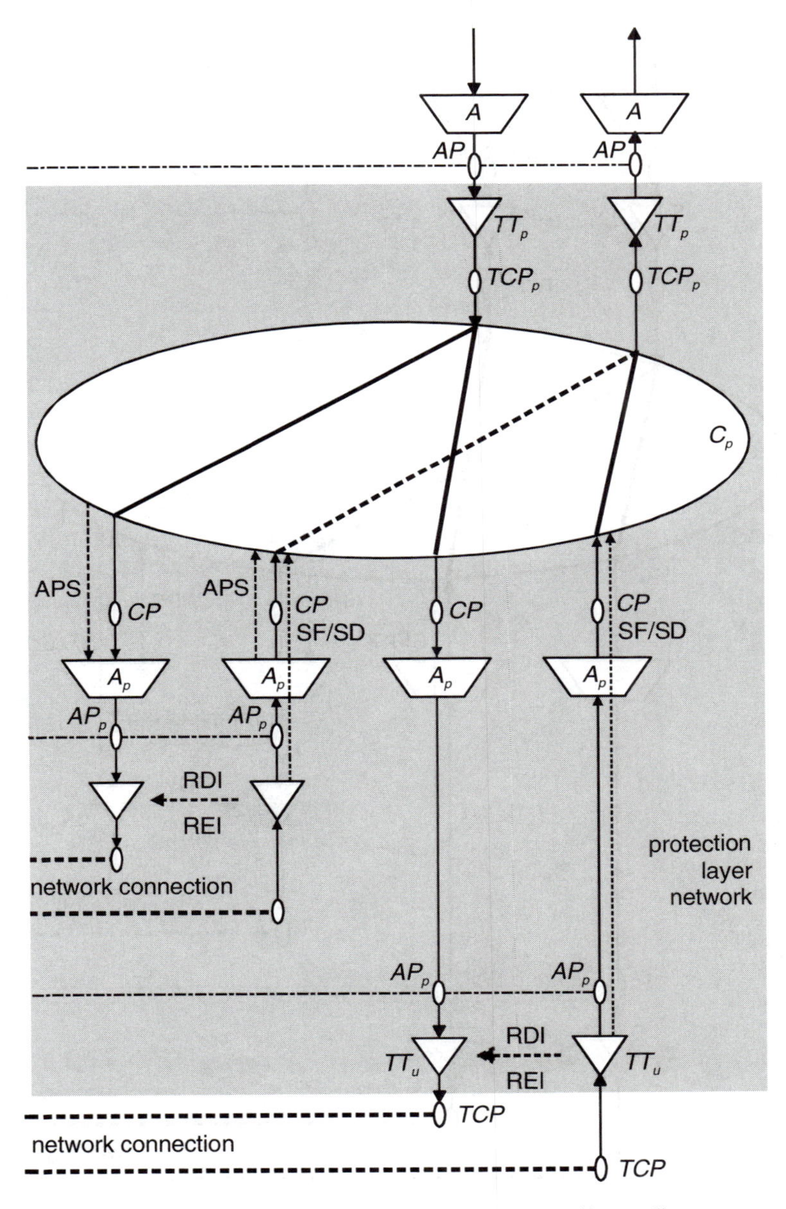

Figure 9.7b 1 + 1 trail protection (far-end).

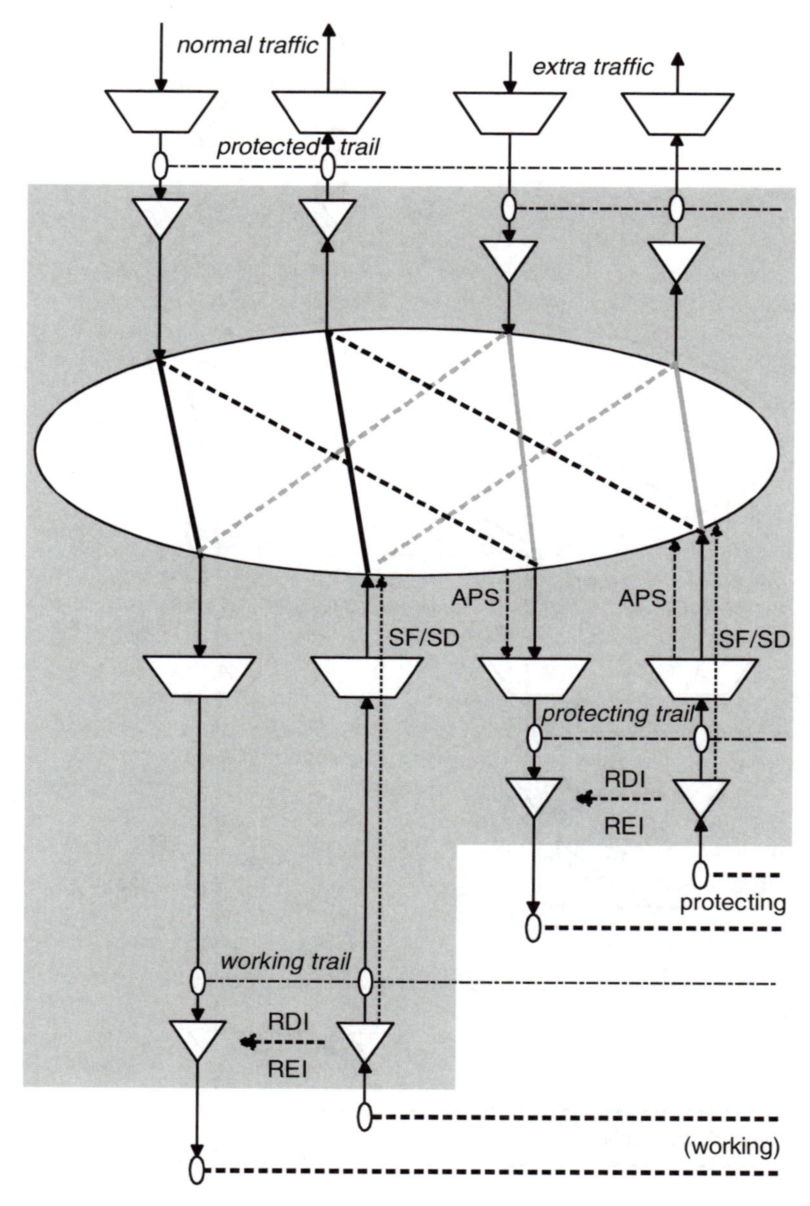

Figure 9.8a 1:1 trail protection (near-end).

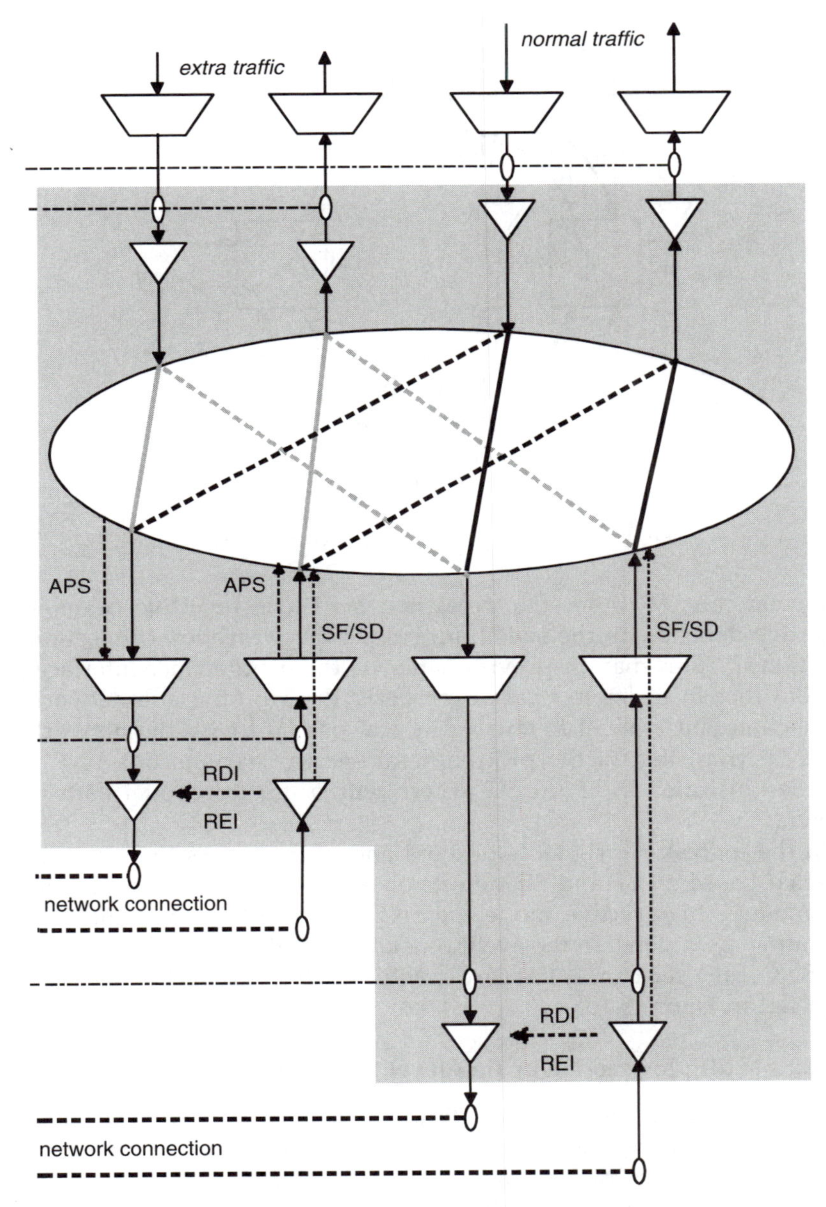

Figure 9.8b 1:1 trail protection (far-end).

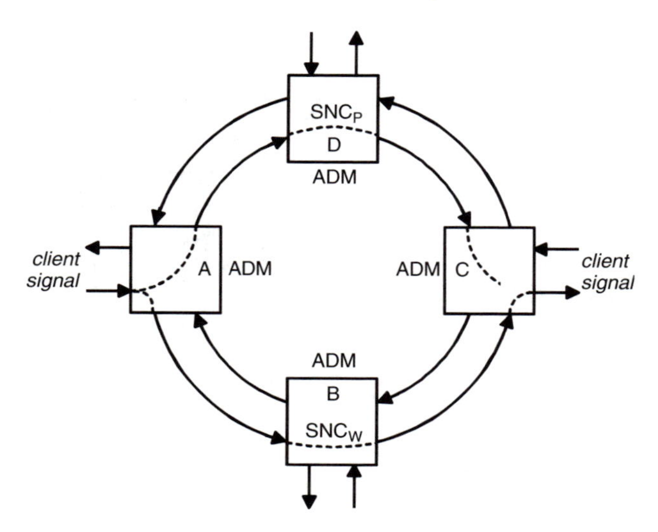

Figure 9.9 Ring under normal conditions.

independently. Whether the protection switch is revertive or non-revertive depends on the switch process in the protection connection function C_P. This may be provisionable via the maintenance interface.

Note that in order to operate properly the working sub-network connection SNC_W shall follow a physical path through the network different from that of the protection sub-network connection SNC_P. This is illustrated in Figure 9.9 where only a uni-directional path is shown.

At the ingress the signal is bridged and at the egress the signal is selected based on SF and SD defects or APS commands according to the models. In revertive mode the SNC_W is selected under normal operating conditions. In the event of a failure in the network affecting the SNC_W the selector at the egress will switch to the SNC_P signal as depicted in Figure 9.10.

9.2.1.2 Multiplex section or line level trail protection

This type of trail protection is applicable only to higher-order network multiplexes that are used to interconnect a number of nodes in a ring configuration, i.e. STM-N with $N = 4, 16, 64, 256$.

The total bandwidth of each section of the ring is shared by the working and protection entities transported over the ring. The bandwidth of the protection entities of each section is shared among the

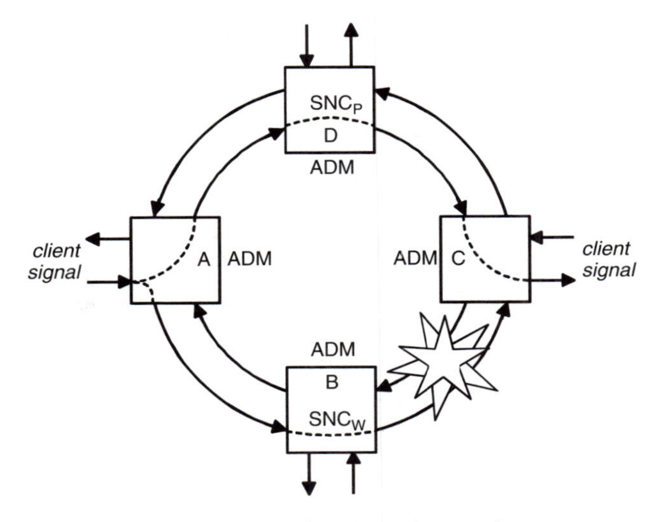

Figure 9.10 Failure in a ring section.

working entities of all sections of the ring. The client signals are transported bi-directionally over the ring and follow the same physical path; together with sharing the protection bandwidth this gives a better utilization of the total bandwidth in the ring. Figure 9.11 shows

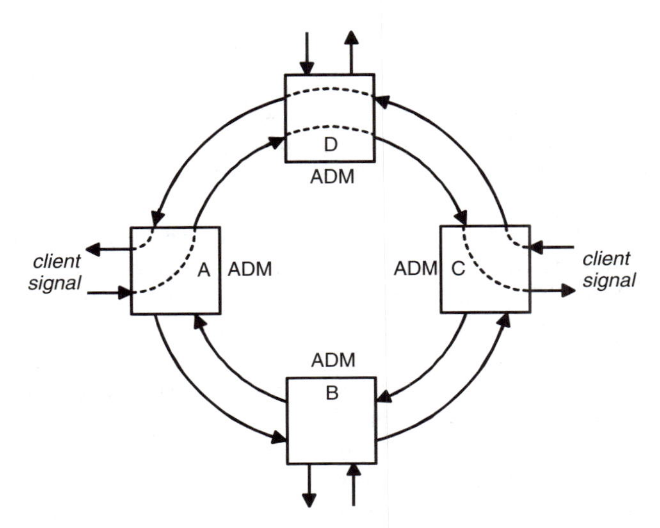

Figure 9.11 Client singal path in MS protection.

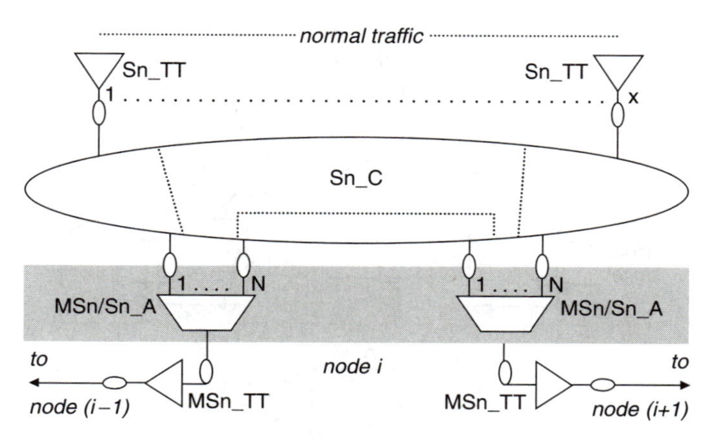

Figure 9.12 Unprotected consideration.

an example of this type of protection in an SDH ring transporting a client signal.

This type of trail protection is also modeled by introducing a protection sublayer. The original section to path adaptation, MSn/Sn_A as shown in the shaded area in Figure 9.12, is expanded according to the rules given in Chapter 4, Section 4.2.1. The result is shown in Figure 9.13.

Note: Figure 9.12 shows a simplified model of node i in a network topology consisting of a ring of nodes; each node is connected to two other nodes. The value of N = 4, 16, 64, 256. The value of x is arbitrary. In the connection function, Sn_C signals are either added to, dropped from or passed through the two multiplex section adaptation functions MSn/Sn_A.

The protection adaptation functions MS_P/Sn_A, the protection trail termination functions MS_P_TT, the protected trail termination functions MS/MS_P_A and an MS protection connection function MS_P are introduced to model the MS – shared protection ring (MS-SPRing) switching. This type of trail protection requires a communication protocol between the control functions of the protection connection functions; the protection adaptation function MS/MS_P_A shall provide access to an Automatic Protection Switch (APS) channel.

With this type of trail protection a distinction is made for the type of traffic that will be transported over the multiplex section between adjacent nodes:

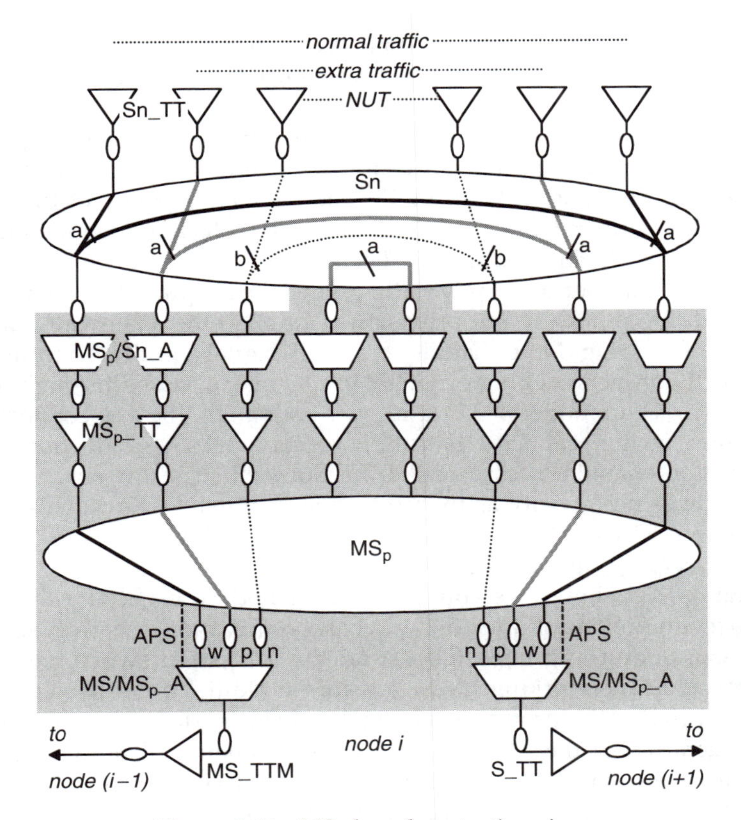

Figure 9.13 MS-shared protection ring.

- normal traffic; this is the traffic than will be protected and will have the highest availability figure. It is transported in the working channel ('w' in Figure 9.13).
- extra traffic; this is the traffic that will be discarded as soon as protection of the normal traffic is required. It will have the lowest availability of the three channels because every protection switch in the ring affects it. The extra traffic is transported in the protection channel ('p' in Figure 9.13).
- non-pre-emptible unprotected traffic (NUT); this is the traffic that will not be affected by the trail protection switching. It will have an availability figure between that of the working channel and the protection channel as it will be affected only by a defect in a multiplex section. The NUT traffic may have its own (trail) protection

mechanism; excluding it will prevent nesting of protection schemes. The NUT channel ('n' in Figure 9.13) carries the NUT.

The maximum payload capacity of the working channel is equal to the capacity of the protection channel, as indicated by 'a' in the Sn connection function in Figure 9.13. The payload capacity of the NUT channel is different from the working and protection channel and is indicated by 'b' in Figure 9.13. The values 'a' and 'b' represent the maximum number of signals Sn that may be transported by a channel. The values 'a' and 'b' are provisioned to have the same value in all nodes of the ring. Note that all 'a' and 'b' signals Sn drawn in the Sn connection function either originate locally or are passed through from node $(i-1)$ to node $(i+1)$ and vice versa. In the Sn connection, function switch capacity has to be reserved to pass through the protection channel in case of a protection switch in the ring. It shall have the same capacity as the protection channel: 'a' Sn signals.

For an STM-N ($N = 4$, 16, 64, 256) the following limits apply: $1 \leq a \leq N/2$, $0 \leq b < N$ and $(2a + b) < N$.

Figure 9.14 shows the configuration of node i, especially the MS_p connection function, after a fault has occurred on the ring section between node i and node $(i+1)$ and the protection switch has been performed. The working traffic towards node $(i+1)$ will replace the extra traffic on the protection channel towards node $(i-1)$; the switch time determines the MTTR of the working channel. The objective is that it is less than 50 ms, see ITU-T Rec. G.841 (1998) clause 7.2.2. All the extra traffic in node i will be interrupted and the NUT to and from node $(i+1)$ will be interrupted affecting their availability; the MTTR is determined by the time it takes to repair physically the faulty section. The equivalent (mirrored) configuration will be found in node $(i+1)$.

Figure 9.15 shows the configuration of node i, especially the MS_p connection function, after a fault has occurred on a ring section *not* connected to node i. To be more precise: if in an STM-N ring one of the sections experiences a fault condition that requires an MS-SPRing protection switch, the two nodes connected to the faulty section will have the switch configuration depicted in Figure 9.14 and all other nodes in the ring will have the configuration depicted in Figure 9.15. In this configuration the protection traffic received from node $(i+1)$ will be passed through to node $(i-1)$ and vice versa using the reserved pass-through capacity of the Sn connection function. All the extra traffic in node i will be interrupted. (Note: the NUT is not affected in this node.)

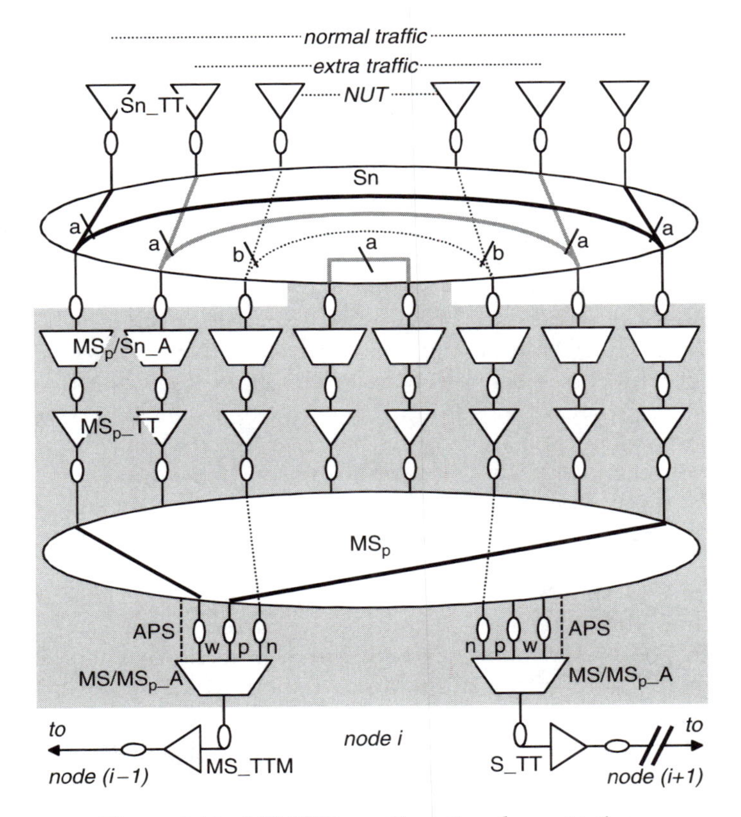

Figure 9.14 MS-SPRing adjacent node protecting.

9.2.2 *Sub-network connection protection*

Sub-Network Connection Protection (SNCP) is a protection mechanism that may be used in any network topology, e.g. mesh, ring or mixed mesh/ring. It is applicable in any path layer in a layered network. It may be used to protect a path between two *Termination Connection Points* (TCP). However, this is useful only when there are two diverse routes available through the network. Figure 9.16 shows a typical SDH network topology: a ring comprised of network elements interconnected by fibers transporting STM-N signals.

At the originating TCP in node i the characteristic information is bridged and sent in both directions on the ring, one signal via node (i − 1) and the other via node (i + 1). At the receiving TCP in node z a

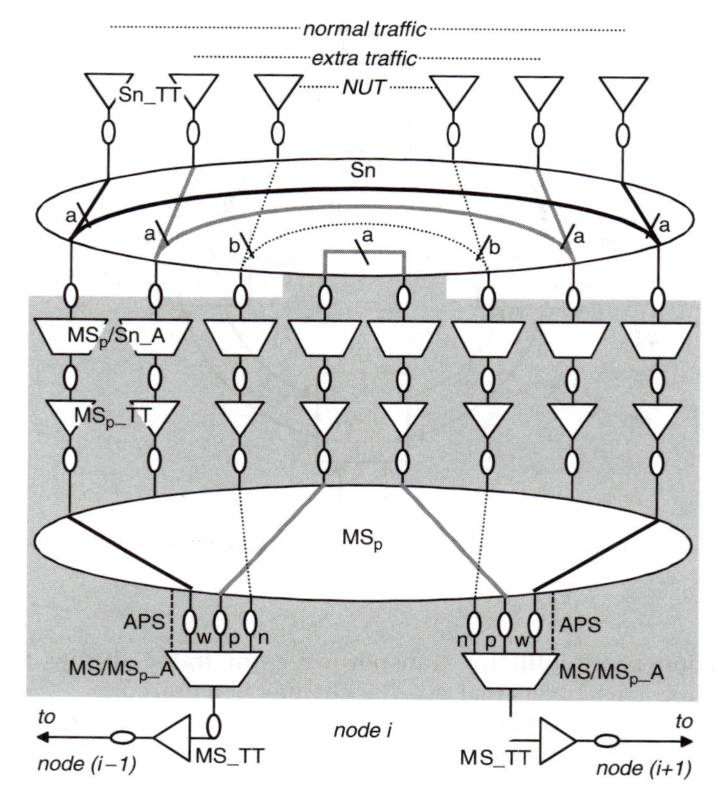

Figure 9.15 MS-SPRing non-adjacent node protecting.

selector is used to select one of these signals based on the performance of either path, e.g. SF or SD.

This protection methodology is also applicable to a part of a path between two *Connections Points* (CP) or between a CP and a TCP provided again that two diverse routes are available. It is even possible that the sub-network connection being protected is made up of a sequence of protected sub-network connections and link connections.

SNCP switching is a methodology that is applied in the client layer network whenever a defect condition is detected in a server layer network, a server sublayer network or another server transport layer network. In other words it is based on server layer performance. Currently, SNCP switching is defined to be uni-directional only; bi-directional SNCP is for further study as well as the use of the

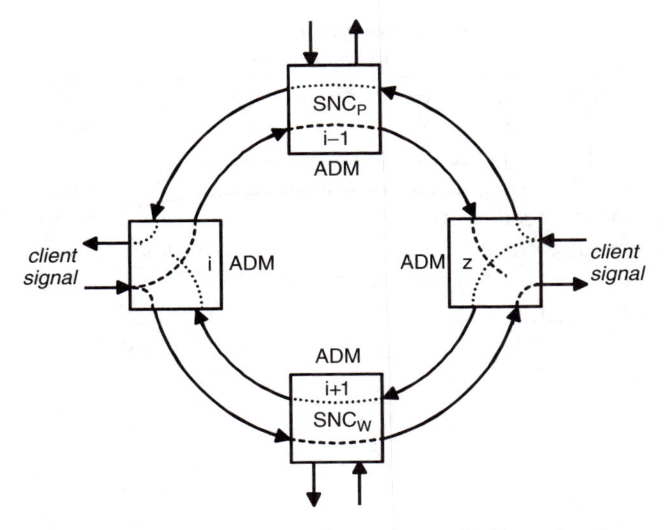

Figure 9.16 Ring with diverse routed client signal.

protection bandwidth for transporting extra traffic during normal operation. This means that no APS channel is required.

SNCP schemes can be characterized by the monitoring method used to derive the switching criteria:

- sub-layer monitoring by expanding the sub-network connection points;
- inherent monitoring by using server layer performance information;
- non-intrusive monitoring by adding client layer monitoring functions.

The signal that is provisioned to be protected, in this case referred to as a working (sub-)network connection, is replaced by a protecting (sub-)network connection if the working (sub-)network connection fails or if its performance falls below the required level.

9.2.2.1 Sublayer trail monitoring:

This SNCP scheme, sometimes referred to as SNC/S, can be modeled by expanding the sub-network connection points and inserting a *protection sublayer*, according to the rules given in Chapter 4, Section 4.2.3. The introduction of a sublayer results in a trail protection of the sublayer trail that runs between the protection access points AP_p. This is illustrated in Figures 9.17a and 9.17b; the inserted protection sublayer is shown in the shaded area. The original connection point is split into a protection connection point CP_p and a protection termination connection point TCP_p. In this inserted sublayer the protection trail termination function TT_p provides the monitoring of the performance of the server signal. (See also Chapter 8, Section 8.2.4.) The monitoring results in a server signal fail (SF) or server signal degraded (SD) signal that is forwarded by the protection adaptation

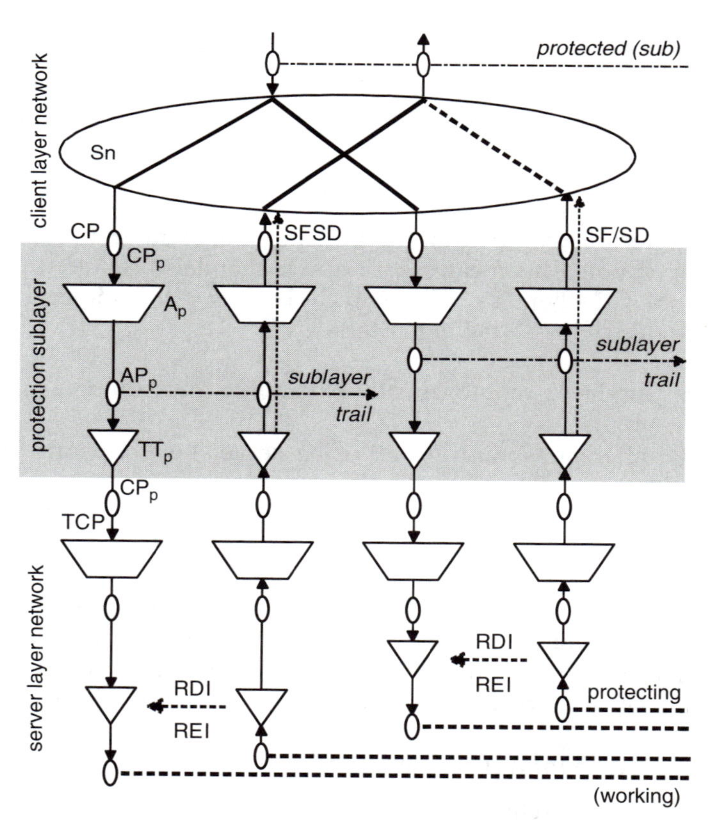

Figure 9.17a SNC protection by sub-layering.

function A_p to the client layer connection function Sn. This Sn will execute the protection switch.

9.2.2.2 Inherent monitoring

This SNCP scheme, normally referred to as SNC/I, can be modeled by showing the server layer performance information derived by the server layer network as described in Chapter 8, Section 8.2.1. The server layer signal condition is used to initiate the protection switch in the client layer. This is illustrated in Figures 9.18a and 9.18b by the shaded area.

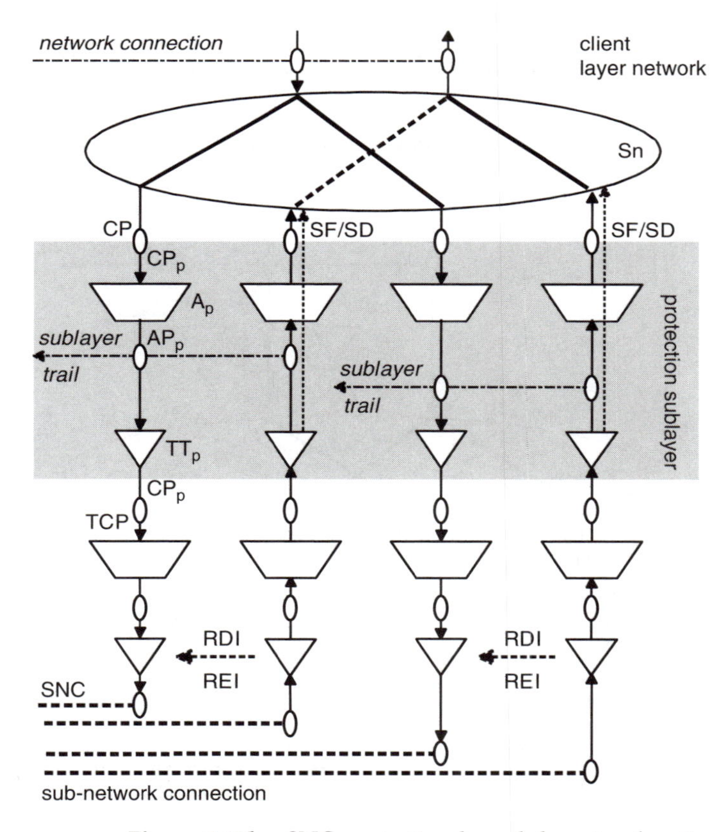

Figure 9.17b SNC protection by sub-layering (continued).

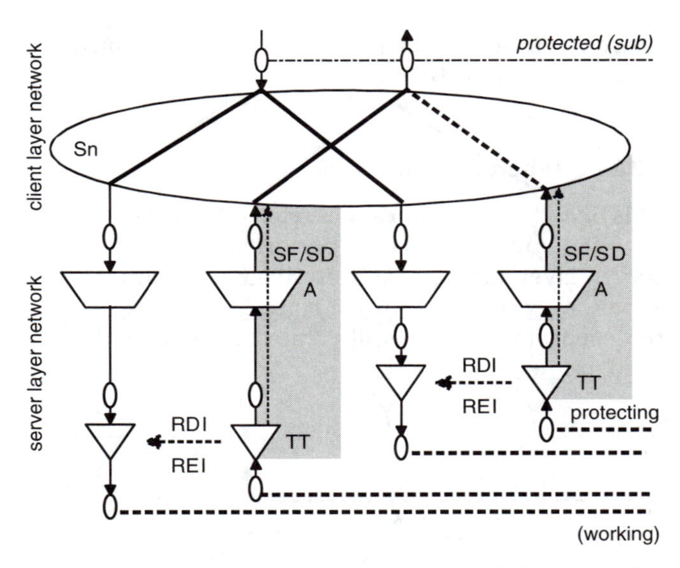

Figure 9.18a SNC protecion by inherent monitoring: SNC/I (continued).

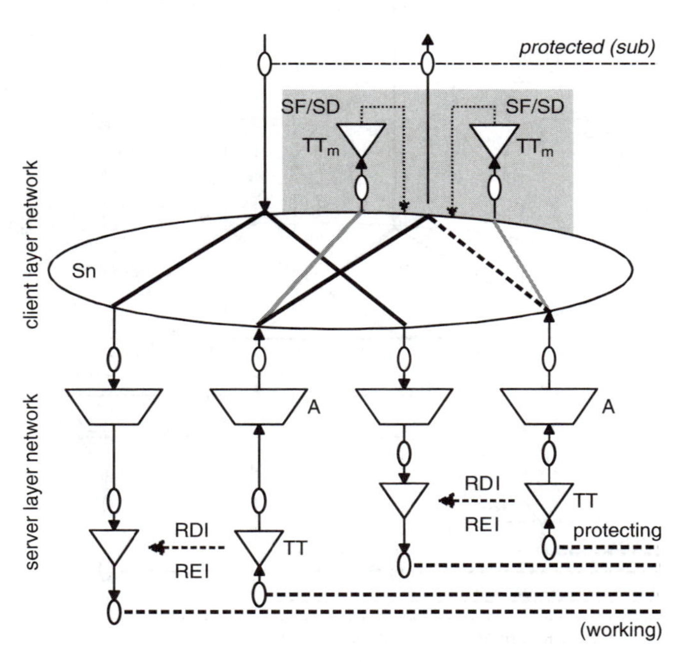

Figure 9.19a SNC protection by non-intrusive monitoring: SNC/N.

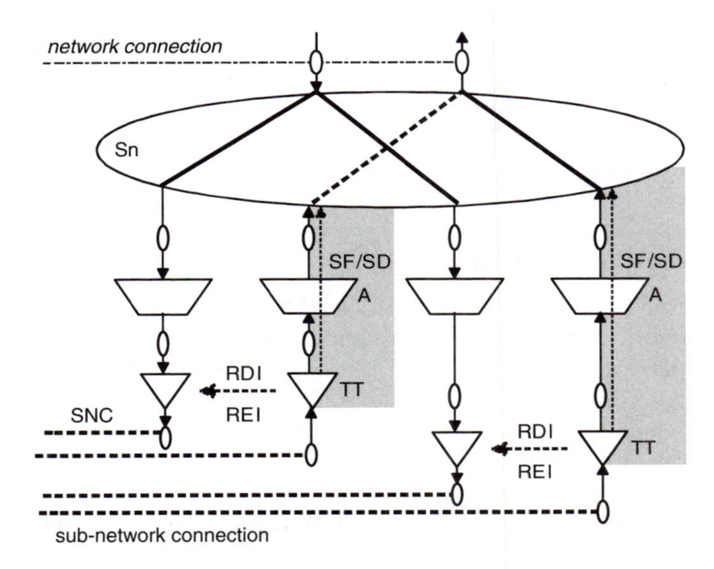

Figure 9.18b SNC protecion by inherent monitoring: SNC/I.

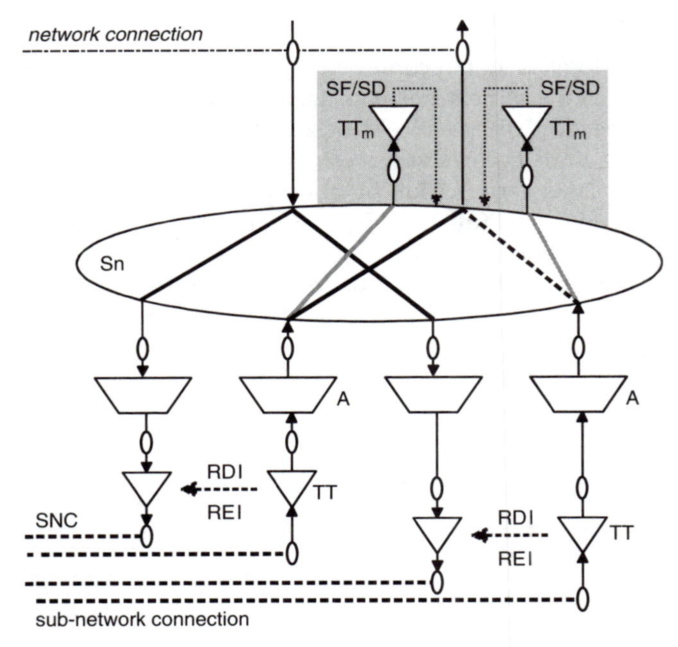

Figure 9.19b SNC protection by non-intrusive monitoring: SNC/N (continued).

The monitoring capabilities of the existing trail termination (TT) is used in the SNC/I scheme. The TT monitors the performance of the server layer (sub-)network connection and detects a server signal fail (SF) or server signal degraded (SD). The SF/SD signal is forwarded by the server layer adaptation function A to the client layer connection function Sn.

9.2.2.3 Non-intrusive monitoring

This SNCP scheme is generally referred to as SNC/N. In this case the server layer network trail termination function (TT) has no capability to monitor the performance of the server layer. For the monitoring of the (sub-)network connection, listen-only trail termination functions TT_m are used. They are available in the client layer (or have to be added as described in Chapter 8, Section 8.2.2) (see the shaded area in Figures 9.19a and 9.19b. These TT_m are connected permanently to the server signal received from the server adaptation function A (see the gray line in the same figures). A detected server signal failure (SF) or server signal degraded (SD) will be reported to the client layer connection function Sn to initiate a protection switch.

9.2.2.4 Intrusive monitoring

This SNCP methodology is not recommended for use as part of a protection scheme. It is the same scheme as used in SNC/N except for the permanent connection of the received server signal. In case of intrusive monitoring, the received server signal is either passed through to the client layer termination function or it is (occasionally) switched to the monitoring function TT_m to get the information required to perform a protection switch. The client signal is interrupted during the monitoring phase. Depending on the frequency of switching for monitoring, this may severely affect the availability of the transported information. See also Chapter 8, Section 8.2.3 for a description of intrusive monitoring.

10

Compound functional models and their decomposition

This chapter describes the compound functional models, and their decompositions. These compound functions were introduced at the time Virtual Concatenation (VCAT) was defined in G.783 (2004). The compound function can be used in the general functional models that describe transport equipment or networks; the implementers of the compound functions can use the decomposed models. Three compound functional models will be described in this chapter: the VCAT compound functions with Link Capacity Adjustment Scheme (LCAS) disabled; the VCAT compound function with LCAS enabled; and the VCAT to Contiguous Concatenation (CCAT) interworking compound function. The VCAT compound function is a specific application of the inverse multiplexing described in Chapter 2, Section 2.5.2.

10.1 LCAS DISABLED VCAT FUNCTIONS

The virtual concatenated trail termination compound function that does not support LCAS or has LCAS disabled is generally referred to as Sn–Xv_TT. It is represented by a compound functional model, which means that it is composed of a number of 'standard' atomic functions. The preferred symbol is depicted in Figure 10.1.

This model can be decomposed into the following atomic functions: the Sn–X Trail Termination function Sn–X_TT, the Sn–Xv to Sn–X

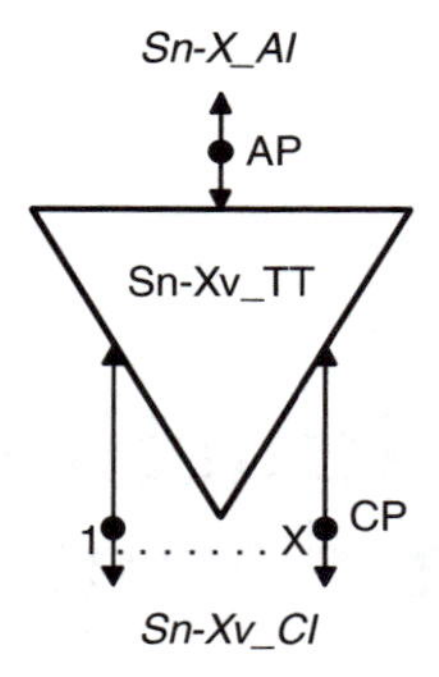

Figure 10.1 Sn–Xv_TT function.

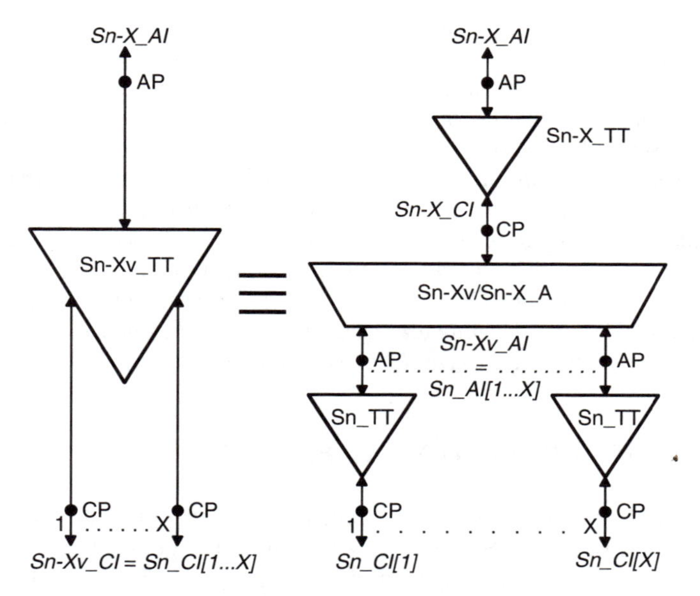

Figure 10.2 Decomposition of the Sn-Xv_TT function.

Adaptation function Sn–Xv/Sn–X_A and the X separate member Sn Trail Termination functions Sn_TT. Figure 10.2 illustrates the decomposition.

This figure shows the bi-directional representation of the atomic functions. The transport of a signal is normally described uni-directionally as follows. A certain trail signal, i.e. the adapted information Sn–X_AI, will be transported through the network using virtual concatenation. For this purpose it is connected at the source side to the source side Sn–X_TT_So function. The source side Sn–Xv/ Sn–X_A_So function takes care that the characteristic information Sn–X_CI, a contiguous signal, received from the Sn–X_TT function will be adapted for distribution over X trails. The adapted information Sn–Xv_AI is equivalent to the sum of the adapted information Sn_AI of the X individual trails. Each source side Sn_TT_So function will output the characteristic information Sn_CI that can be transported individually through the network. In the opposite direction, the Sn_CI signal received by the X individual sink side Sn_TT_Sk functions will be processed and sent to the sink side Sn–Xv/Sn–X_A_Sk function to compose the concatenated characteristic information Sn–X_CI that will be processed by the sink side Sn–X_TT_Sk function and output as the trail signal Sn–X_AI.

The following sections will describe the individual atomic functions.

10.1.1 Sn–Xv trail termination function

This is the generic virtual concatenation trail termination function that does not support LCAS (or has LCAS disabled). It serves as the beginning and end point of a contiguous trail signal that will be transported over X individual (sub-)network connections. Figure 10.3 shows the bi-directional model.

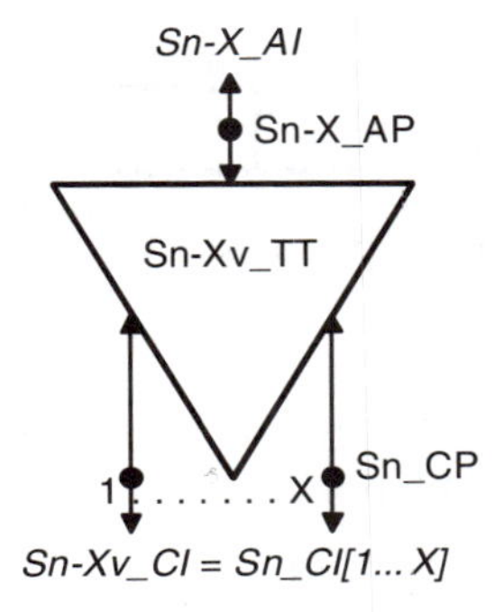

Figure 10.3 Bi-directional Sn-Xv_TT function.

Bi-directional model Sn–Xv_TT

The *Adapted Information* Sn–X_AI present at the *Access point* Sn–X_AP is adapted for the transport over the network and distributed over the X individual *Characteristic Information* Sn_CI, passing the *Connection Points* Sn_CP. The operation of the function is described in the next sections of this chapter.

This model can be decomposed in the uni-directional models shown in Figures 10.4 and 10.5.

Source side: Sn–X_TT_So

Symbol

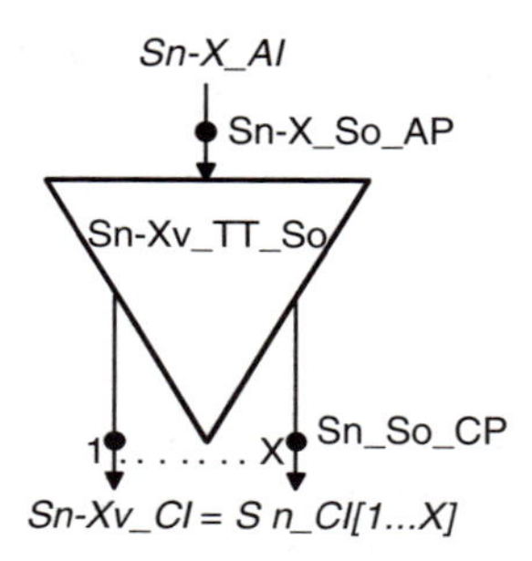

Figure 10.4 Source side Sn-Xv_TT function.

Interfaces

Sn–Xv_TT_So.

inputs	outputs
at Sn–X_So_AP Sn–X_AI_D Sn–X_AI_CK Sn–X_AI_FS	at Sn_So_CP Sn_CI[1...X]_D Sn_CI[1...X]_CK Sn_CI[1...X]_FS

Processes

See the description of the individual atomic functions in the following sections.

Sink side: Sn–X_TT_Sk

Symbol

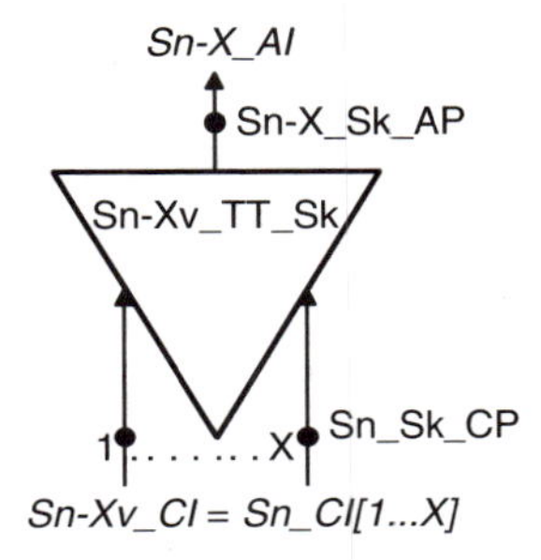

Figure 10.5 Sink side Sn-Xv_TT function.

Interfaces

Sn–Xv_TT_Sk.

inputs	outputs
at Sn–X_Sk_CP Sn–X_CI_D Sn–X_CI_CK Sn–X_CI_FS	at Sn_Sk_AP Sn_AI[1...X]_D Sn_AI[1...X]_CK Sn_AI[1...X]_FS

Processes

See the description of the individual atomic functions in the following sections.

10.1.2 Sn–Xv/Sn–X adaptation function

This is the function that performs the adaptation of the contiguous client signal to the virtual concatenated server layer signals transported by the individual members of the *Virtual Concatenation Group* (VCG). Figure 10.6 shows the bi-directional functional model.

Bi-directional model: Sn–Xv/Sn–X_A

Symbol

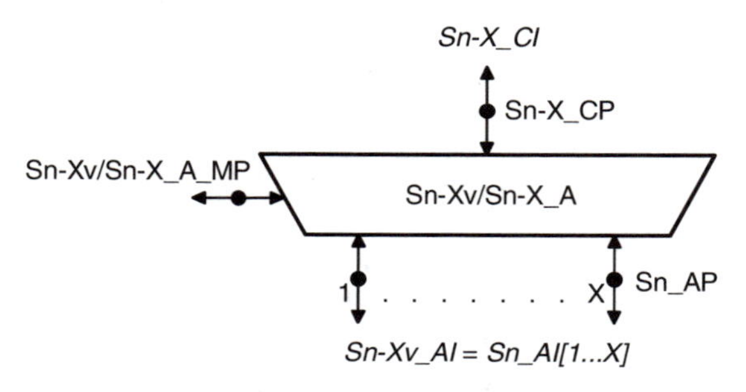

Figure 10.6 Bi-directional Sn-Xv/Sn-X_A function.

This model can be further decomposed into the uni-directional models depicted in Figures 10.7 and 10.9.

Source side: Sn–Xv/Sn–X_A_So

The source side VCAT adaptation function performs the distribution of the incoming contiguous concatenated signal Sn–Xc received at the Connection Point as data signal Sn–X_CI_D (together with the associated clock Sn–X_CI_CK and frame start signal Sn–X_CI_FS) to constitute the virtual concatenated signal Sn–Xv transmitted at the Access Point as data signal Sn–Xv_AI_D. This signal is the equivalent of the combined payload of the X individual member signals Sn transmitted as Sn_AI_D over the member Access Points, where X can have any value $1 \leq X \leq (max)$ (where (max) is technology specific, e.g. 256 for a VC–4).

Symbol

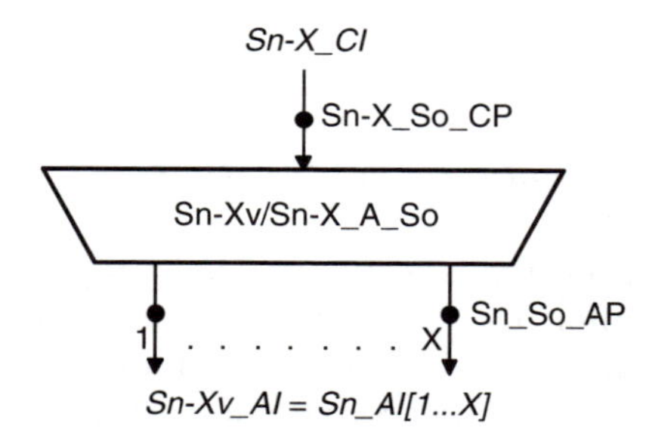

Figure 10.7 Source side Sn-Xv/Sn-X_A function.

Interfaces

Sn–Xv/Sn–X_A_So.

inputs	outputs
at Sn–X_So_CP	**at Sn_So_AP**
Sn–X_CI_D	Sn_AI[1...X]_D
Sn–X_CI_CK	Sn_AI[1...X]_CK
Sn–X_CI_FS	Sn_AI[1...X]_FS

Processes

This function performs the following processing between its input and output:

- distributes the received client signal round-robin octet-wise over the payload area of all the members in the VCG;
- inserts the VCAT overhead information in each of the VCG member signals, i.e. the multi-frame indication, sequence number and CRC;
- inserts the payload type identification into the signal label of each of the member signals.

The processes described above and their relations are depicted in Figure 10.8.

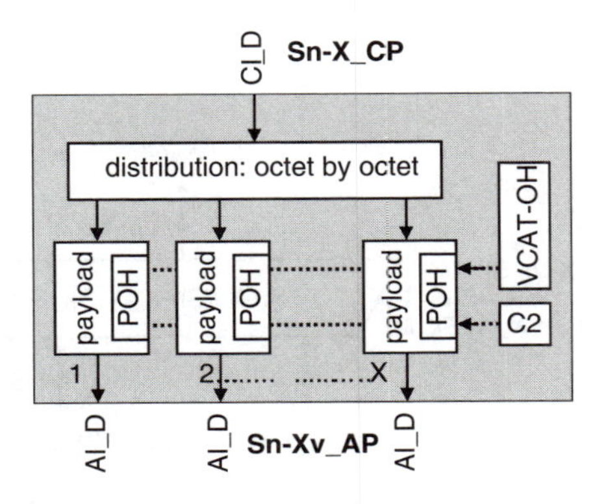

Figure 10.8 Sn-Xv/Sn-X_A So processes.

Defects

None.

Consequent actions

None.

Defect correlation

None.

Performance monitoring

None.

Sink side: Sn–Xv/Sn–X_A_Sk

The sink side VCAT adaptation function performs the reconstruction of the contiguous concatenated signal Sn–Xc transmitted at the Connection Point as Sn–X_CI_D from the virtual concatenated signal Sn–Xv received at the Access Point as Sn–Xv_AI_D. This signal is the equivalent of the combined payload of the X individual member signals Sn received as Sn_AI_D at the member Access Points, where X can have any value $1 \leq X \leq (max)$ (where (max) is technology specific, e.g. 256 for a VC–4).

Symbol

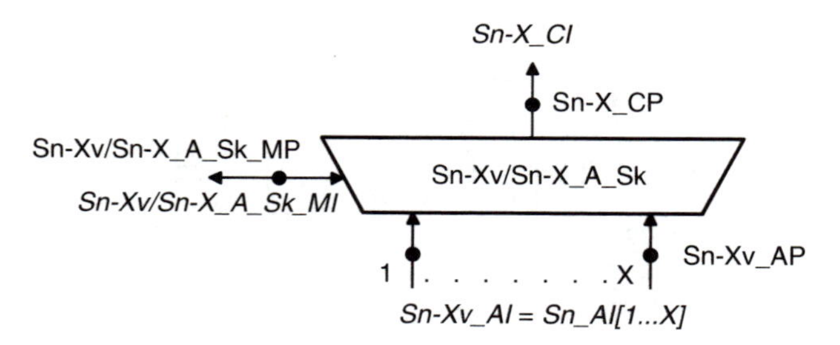

Figure 10.9 Sink side Sn-Xv/Sn-X_A function.

Interfaces

Sn–Xv/Sn–X_A_Sk

inputs	outputs
at Sn_Sk_AP	**at Sn_Sk_CP**
Sn_AI[1...X]_D	Sn–X_CI_D
Sn_AI[1...X]_CK	Sn–X_CI_CK
Sn_AI[1...X]_FS	Sn–X_CI_FS
Sn_AI[1...X]_TSF	Sn–X_CI_SSF
	at Sn–Xv/Sn–X_A_Sk_MP
	Sn–Xv/Sn–X_A_Sk_MI_cPLM[1...X]
	Sn–Xv/Sn–X_A_Sk_MI_AcSL[1...X]
	Sn–Xv/Sn–X_A_Sk_MI_AcSQ[1...X]
	Sn–Xv/Sn–X_A_Sk_MI_cLOM[1...X]
	Sn–Xv/Sn–X_A_Sk_MI_cSQM[1...X]
	Sn–Xv/Sn–X_A_Sk_MI_cLOA

Processes

This function performs the following processing between its input and output:

- verifies the payload type identification in the signal label of each of the member signals;
- buffers the received signal of each individual VCG member to compensate the differential delay due to diverse routing;
- assembles the original client payload signal from the member octets after using the multi-frame indication for re-alignment and the sequence number for re-sequencing the individual member signals;
- forwards a received *Trail Signal Fail* (TSF) indication to the next function as *Server Signal Fail* (SSF).

The processes described above and their relations are depicted in Figure 10.10.

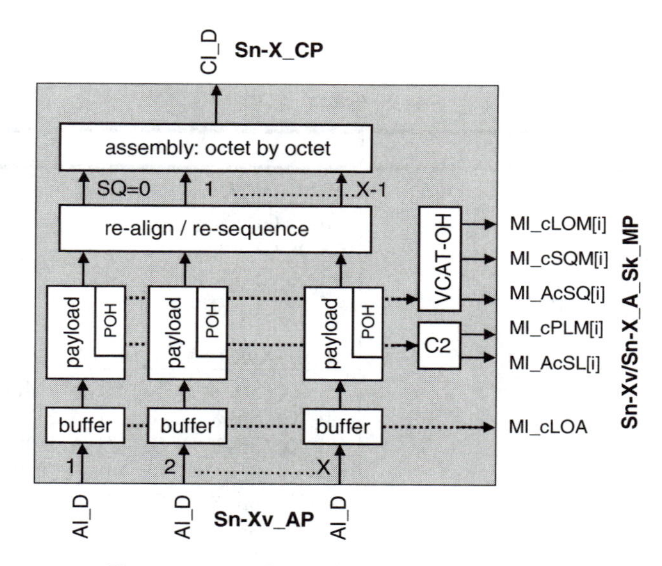

Figure 10.10 Sn-Xv/Sn-X_A Sk processes.

Defects

The following defects will be detected by this function:

- *PayLoad Mismatch* (PLM): The received signal label does not match the value of the payload type specific for this function. The function reports the received and *Accepted Signal Label* (AcSL).
- *Loss of Multi-frame* (LOM): The VCAT overhead multi-frame cannot be recovered from the received signal.
- *Sequence number Mismatch* (SQM): The received SQ number of a member is in conflict with the SQ of other members in the VCG. The function reports the received and *Accepted SQ number* (AcSQ).
- *Loss of Alignment* (LOA): The differential delay, i.e. the difference in time between the member experiencing the least propagation delay and the member experiencing the most propagation delay in the network, exceeds the implemented alignment buffer.

Consequent actions

Upon detection of one or more of the defects, the consequent action is to replace the outgoing signal by an all-ONEs signal, i.e. the *Alarm Inhibit Signal* (AIS), and sending the SSF indication to the next function.

Defect correlation

When the defects are persistent the correlation process determines the most probable cause and reports this via the Sn–Xv/Sn–X_A_Sk_MP management point.

Performance monitoring

None.

10.1.3 *Sn–X trail termination function*

This function serves as the start and end point of the virtual trail that is used to transport the virtual concatenated signal through the network. The bi-directional model is shown in Figure 10.11.

Bi-directional model: Sn–X_TT

Symbol

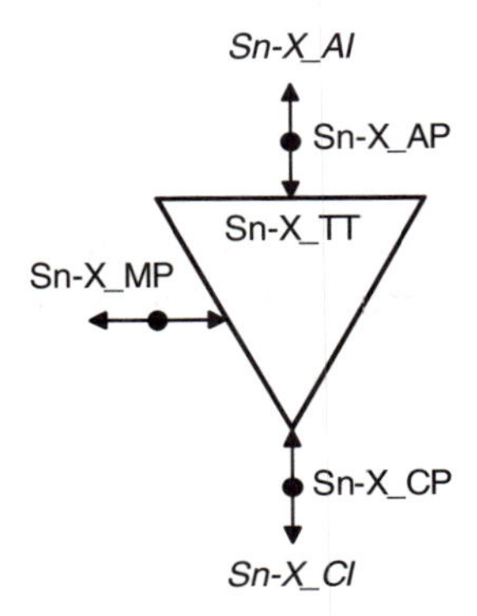

Figure 10.11 Bi-directional Sn-X_TT function.

This model can be further decomposed into the uni-directional source and sink models illustrated in Figures 10.12 and 10.13.

Source side: Sn–X_TT_So

Symbol

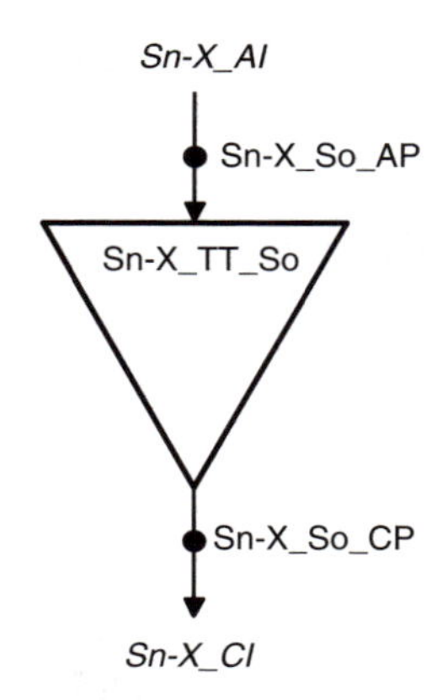

Figure 10.12 Source side Sn-X_TT function.

Interfaces

Sn–X_TT_So.

inputs	outputs
at Sn–X_So_AP	**at Sn–X_So_CP**
Sn–X_AI_D	Sn_CI_D
Sn–X_AI_CK	Sn_CI_CK
Sn–X_AI_FS	Sn_CI_FS

Processes

The Sn–Xv trail termination function at the source side is used only for providing a correct model. It does not contain any specific processes.

Defects

None.

Consequent actions

None.

Defect correlation

None.

Performance monitoring

None.

Sink side: Sn–X_TT_Sk

Symbol

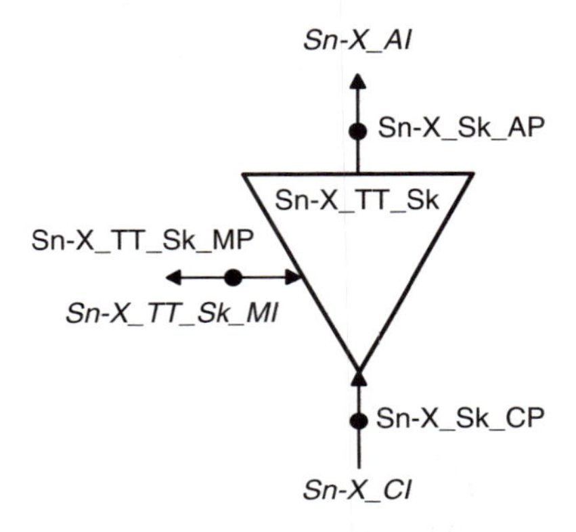

Figure 10.13 Sink side Sn-X_TT function.

Interfaces

Sn–X_TT_Sk.

inputs	outputs
at Sn–X_Sk_CP Sn–X_CI_D Sn–X_CI_CK Sn–X_CI_FS Sn–X_CI_SSF	**at Sn_Sk_AP** Sn_AI[1...X]_D Sn_AI[1...X]_CK Sn_AI[1...X]_FS Sn_AI[1...X]_TSF
at Sn–X_TT_Sk_MP Sn–X_TT_Sk_MI_SSF_Reported	**at Sn–X_TT_Sk_MP** Sn–X_TT_Sk_MI_cSSF

Processes

The Sn–Xv trail termination function at the sink side transfers the concatenated signal without further processing; it is used only for correct modeling.

Defects

None.

Consequent actions

None.

Defect correlation

A received CI_SSF will be forwarded as AI_TSF to the next function to maintain the transparency. If the SSF reporting has been enabled (MI_SSF_Reported = true), the cSSF condition will be reported via the management point Sn–X_TT_Sk_MP.

Performance monitoring

None.

10.1.4 Sn trail termination function

This function serves as the start and end point of each of the X member trails that are actually used to transport the virtual concatenated signal through the network. These are the trails that intermediate nodes will 'see' without knowing that they are members of a VCG. Figure 10.14 shows the bi-directional model.

Bi-directional model: Sn_TT

Symbol

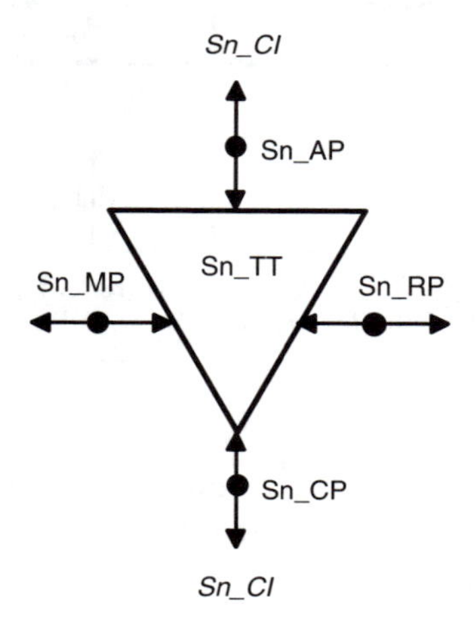

Figure 10.14 Bi-directional Sn_TT function.

This function is described in detail in Chapter 6, Section 6.2.1.

10.2 LCAS-CAPABLE VCAT FUNCTIONS

The LCAS-capable virtual concatenated Sn path layer functional model is similar to the virtual concatenation model with the addition of the LCAS functionality. To make the distinction it will be referred to as Sn–Xv–L.

10.2.1 Sn–Xv–L layer trail termination function

The decomposition for this virtual concatenation function with LCAS enabled is almost the same as the one shown and described in Section 10.1 of this chapter. The Sn–Xv–L_TT function can be further decomposed as depicted in Figure 10.15.

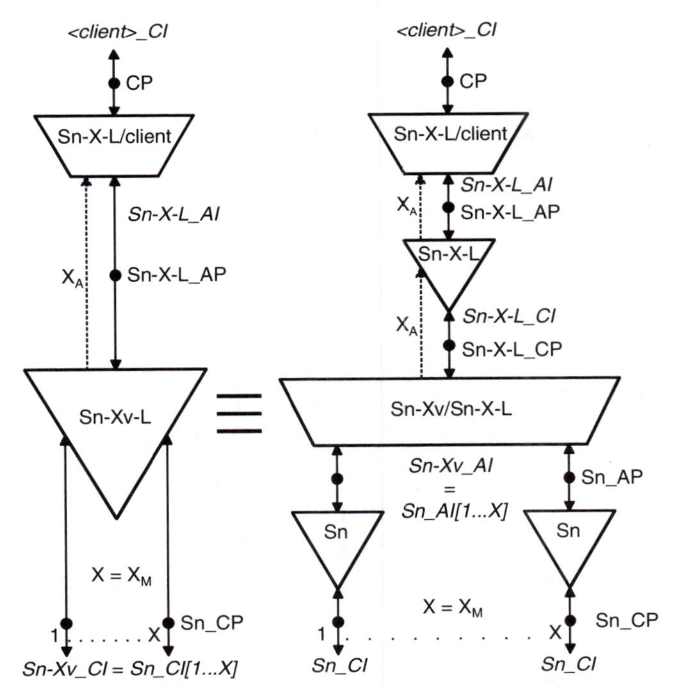

Figure 10.15 Decomposition of Sn-Xv-L_TT function.

Note the following technology-specific particularizations:

- The Sn–Xv/Sn–X–L_A adaptation function contains LCAS specific processes.
- The Sn–X–L_TT termination function contains LCAS specific processes.
- The maximum value of X, i.e. X_M, or X_{MT} for the transmit side and X_{MR} for the receive side, is technology specific, e.g. X = 64 for S12. The number of members that are actually carrying payload or the active members is X_A.

10.2.2 *Sn–Xv/Sn–X–L adaptation function*

This is the function that performs the adaptation of the (contiguous concatenated) client signal to the (virtual concatenated) server layer signals transported by the members of the VCG. An indication of the available transport bandwidth (X_A) is sent towards the client layer. Figure 10.16 shows the bi-directional model.

Bi-directional model: Sn–Xv/Sn–X–L_A

Symbol

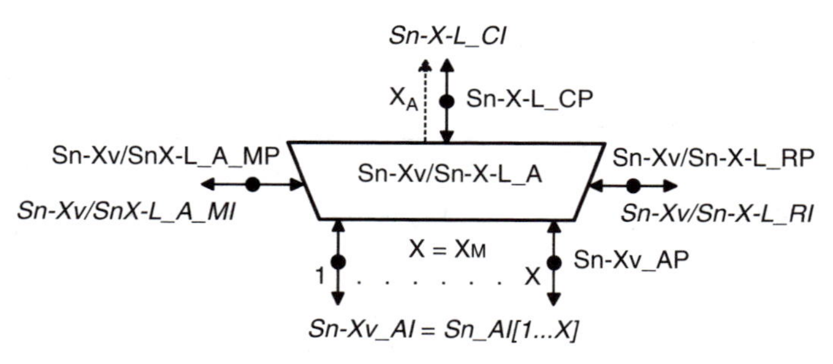

Figure 10.16 Bi-directional Sn-Xv/Sn-X-L_A function.

This model can be further decomposed into the uni-directional source and sink atomic models (Figures 10.17 and 10.19).

Source side: Sn–Xv/Sn–X–L_A_So

Symbol

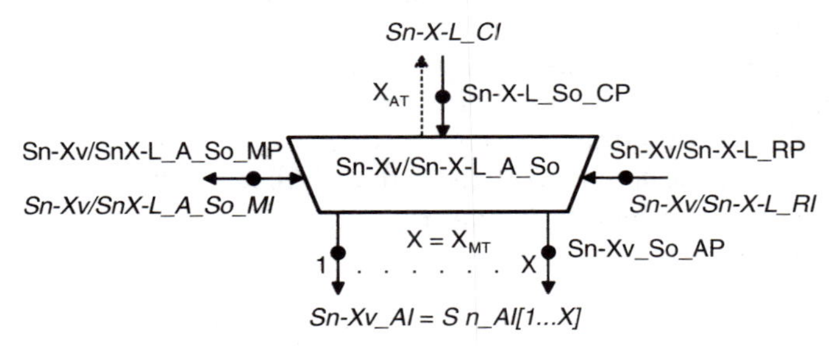

Figure 10.17 Source side Sn-Xv/Sn-X-L_A function.

Interfaces

Sn–Xv/Sn–X–L_A_So.

inputs	outputs
at Sn–X–L_So_CP Sn–X–L_CI_D Sn–X–L_CI_CK Sn–X–L_CI_FS	**at Sn–X–L_So_CP** Sn–X–L_CI_X_{AT} **at Sn–Xv_So_AP** Sn_AI[1...X]_D Sn_AI[1...X]_CK Sn_AI[1...X]_FS
at Sn–Xv/Sn–X–L_A_So_MP Sn–Xv/Sn–X–L_A_So_MI_LCASEnable Sn–Xv/Sn–X–L_A_So_MI_ProvM[1...X] Sn–Xv/Sn–X–L_A_So_MI_PLCTThr	**at Sn–Xv/Sn–X–L_A_So_MP** Sn–Xv/Sn–X–L_A_So_MI_X_{AT} Sn–Xv/Sn–X–L_A_So_MI_X_{MT} Sn–Xv/Sn–X–L_A_So_MI_TxSQ[1...X] Sn–Xv/Sn–X–L_A_So_MI_cPLCT Sn–Xv/Sn–X–L_A_So_MI_cTLCT Sn–Xv/Sn–X–L_A_So_MI_cFOPT
at Sn–Xv/Sn–X–L_RP Sn–Xv/Sn–X–L_RI_RS_Ack_rec Sn–Xv/Sn–X–L_RI_RS_Ack_gen Sn–Xv/Sn–X–L_RI_MST_rec[0...(max)] Sn–Xv/Sn–X–L_RI_MST_gen[0...(max)]	

Processes

The process definitions for this function are the same as for the *Sn–Xv/Sn–X_A_So* function described in Section 10.1.2, when the LCAS capability is enabled (MI_LCASEnable = true), with the following technology-specific particularizations:

- determines whether the payload size of the VCG matches the size of the client signal;
- determines the provisioned size X_{PT} of the VCG and the maximum VCG size X_{MT}; each member is provisioned (MI_ProvM[i]) to be a member of the group;
- determines the actual size X_{AT} of the VCG; the number of active, i.e. payload transporting, members;
- determines which members are (temporarily) failed;
- provides a temporary VCG size reduction, and hitless addition of a previously failed member in coordination with the sink function;
- provides a hitless increase and decrease of the VCG X_{AT}, coordinated with the sink function;
- inserts the VCAT overhead information in each of the VCG member signals, i.e. VCG identification, multi-frame indication, sequence number, acknowledge, member status and CRC; the SQ number assigned to each member is available as TcSQ at the management point.

A detailed description of the processes is provided in ITU-T Rec. G.806 (2004) clause 10.1. The processes described above and their relations are depicted in Figure 10.18.

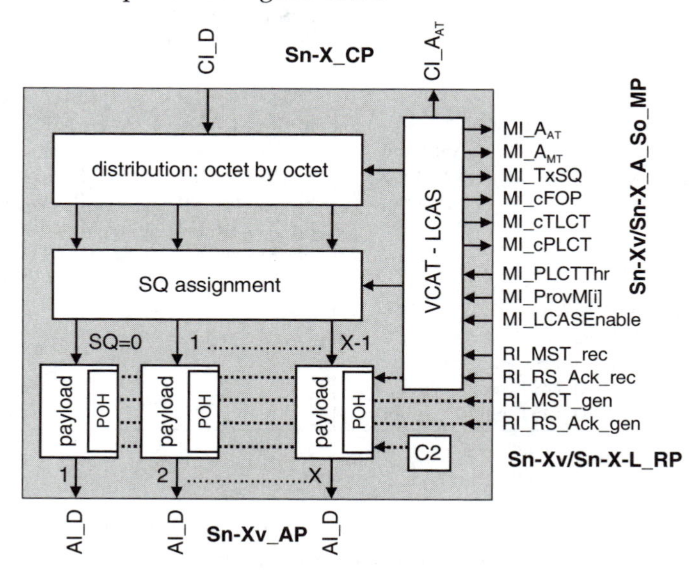

Figure 10.18 Sn-Xv/Sn-X-L_A_So processes.

Defects

Unexpected persistent member status MST shall be detected.

Consequent actions

None.

Defect correlation

The following probable causes are detected and reported by the distribution process:

- If the LCAS procedure receives unexpected events the process may be in the wrong state; this is reported as *Failure Of Protocol Transmit* (FOPT).
- If one or more members is removed temporarily from the VCG due to a network problem, a *Partial Loss of payload Capacity Transmit* (PLCT) may be reported; a threshold (MI_PLCTThr) can be set.
- If all members in the VCG are removed temporarily, a *Total Loss of payload Capacity Transmit* (TLCT) is reported.

Performance monitoring

None.

Sink side: *Sn–Xv/Sn–X–L_A_Sk*

Symbol

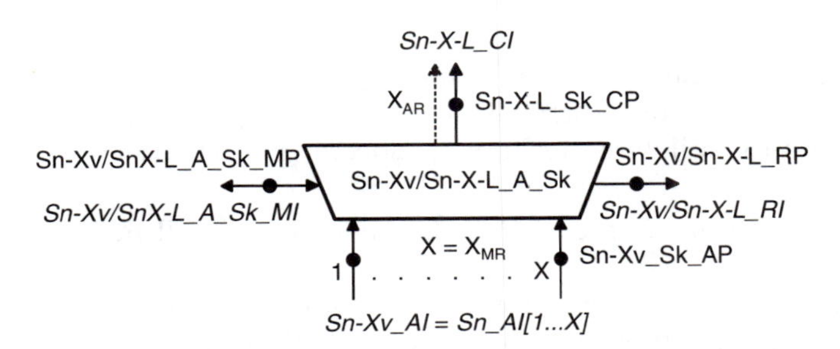

Figure 10.19 Sink side Sn-Xv/Sn-X-L_A function.

Interfaces

Sn–Xv/Sn–X–L_A_Sk.

inputs	outputs
at Sn–Xv_Sk_AP	**at Sn–X–L_Sk_CP**
Sn_AI[1...X]_D	Sn–X–L_CI_D
Sn_AI[1...X]_CK	Sn–X–L_CI_CK
Sn_AI[1...X]_FS	Sn–X–L_CI_FS
Sn_AI[1...X]_TSF	Sn–X–L_CI_SSF
Sn_AI[1...X]_TSD	Sn–X–L-CI_X_{AR}
at Sn–Xv/Sn–X–L_A_Sk_MP	**at Sn–Xv/Sn–X–L_A_Sk_MP**
Sn–Xv/Sn–X–L_A_Sk_MI_ProvM[1...X]	Sn–Xv/Sn–X–L_A_Sk_MI_X_{MR}
Sn–Xv/Sn–X–L_A_Sk_MI_LCASEnable	Sn–Xv/Sn–X–L_A_Sk_MI_X_{AR}
Sn–Xv/Sn–X–L_A_Sk_MI_PLCRThr	Sn–Xv/Sn–X–L_A_Sk_MI_DMFI[1...M]
Sn–Xv/Sn–X–L_A_Sk_MI_TSDEnable	Sn–Xv/Sn–X–L_A_Sk_MI_LCAS_So_Detected
Sn–Xv/Sn–X–L_A_Sk_MI_HOTime	Sn–Xv/Sn–X–L_A_Sk_MI_cPLCR
Sn–Xv/Sn–X–L_A_Sk_MI_WTRTime	Sn–Xv/Sn–X–L_A_Sk_MI_cTLCR
	Sn–Xv/Sn–X–L_A_Sk_MI_cFOPR
	Sn–Xv/Sn–X–L_A_Sk_MI_cLOM[1...X]
	Sn–Xv/Sn–X–L_A_Sk_MI_cSQM[1...M]
	Sn–Xv/Sn–X–L_A_Sk_MI_cMND[1...M]
	Sn–Xv/Sn–X–L_A_Sk_MI_AcSQ[1...M]
	Sn–Xv/Sn–X–L_A_Sk_MI_cPLM[1...M]
	Sn–Xv/Sn–X–L_A_Sk_MI_AcSL[1...M]
	at Sn–Xv/Sn–X–L_RP
	Sn–Xv/Sn–X–L_RI_RS_Ack_rec
	Sn–Xv/Sn–X–L_RI_RS_Ack_gen
	Sn–Xv/Sn–X–L_RI_MST_rec[0...(max)]
	Sn–Xv/Sn–X–L_RI_MST_gen[0...(max)]

Processes

The process definitions for this function are the same as for the *Sn–Xv/ Sn–X_A_Sk* function described in Section 10.1.2, if MI_LCASEnable = true, with the following technology-specific particularizations:

- determines if the received VCG payload size will fit in the client signal;
- determines the provisioned size X_{PR} and maximum size X_{MR} of the VCG; each member is provisioned (MI_ProvM[i]) to be a member of the group;

- determines the actual size X_{AR} of the VCG; the number of active, i.e. payload transporting, members;
- determines which VCG members are (temporarily) failed;
- provides a temporary group size reduction, and supports hitless addition of a previously failed signal in coordination with the source function. The reduction may be based on degraded signals if MI_TSDEnable = true; the reduction can be postponed (hold-off) by time MI_HOTime and the addition (wait to restore) by time MI_WTRtime;
- provides a hitless increase and decrease of the active group X_{AR}, coordinated with the source function.

A detailed description of the processes is provided in ITU-T Rec. G.806 (2004) clause 10.1. The processes described above and their relations are depicted in Figure 10.20.

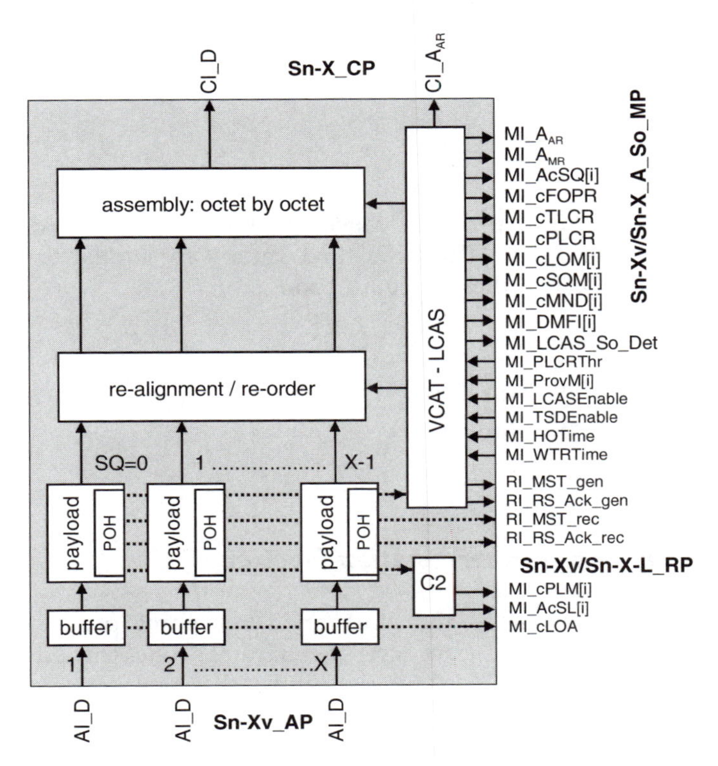

Figure 10.20 Sn-Xv/Sn-X-L_A_Sk processes.

Defects

The following defects will be detected and reported:

- *Loss of Multi-frame* (dLOM) if the LCAS overhead cannot be recovered or aligned between members;
- *Inconsistent sequence numbering* (dSQNC);
- *Member Not De-skewable* (dMND) if the differential delay is too large for the buffers;
- Persistent CRC errors (dCRC).

Consequent actions

The consequent action is the maintenance signal *Member Service Unavailable—LCAS active* (mMSU–L) if either a signal fail is received from a member Sn_TT, or a LOM or an MND is detected. Consequently, the output signal is replaced by an AIS signal together with an SSF indication.

Defect correlation

Determines the most probable causes cLOM, cMND, cLOA, cSQM, cPLCR, cTLCR, cFOPR. The latter three causes are similar to cPLCT, cTLCT and cFOPT in the source function.

Performance monitoring

None.

10.2.3 Sn–X–L trail termination function

This function serves as the start and end point of the virtual trail that is used to transport the virtual concatenated signal through the network. If LCAS is enabled the payload size of the virtual trail can vary. The bi-directional model is shown in Figure 10.21.

Bi-directional model: Sn–Xv/Sn–X_So_A

Symbol

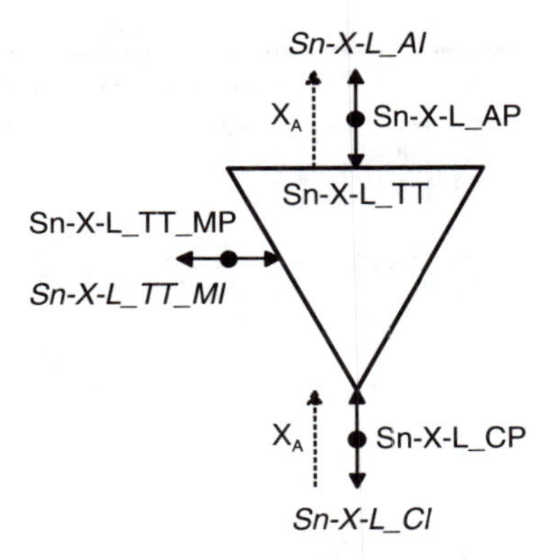

Figure 10.21 Bi-directional Sn-X-L_TT function.

This model can be further decomposed into the uni-directional source and sink models (see Figures 10.22 and 10.23).

Source side: Sn–X–L_TT_So

Symbol

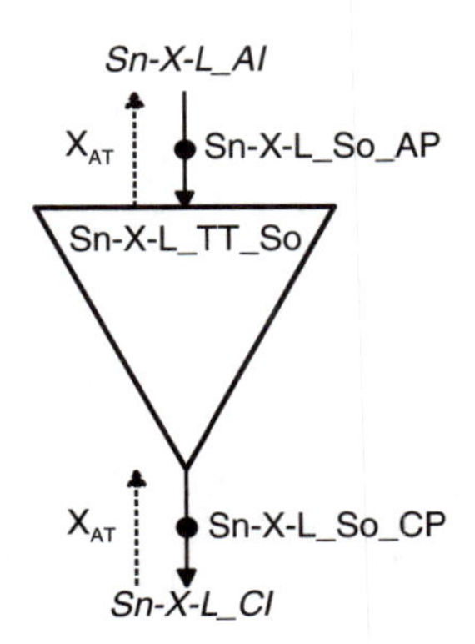

Figure 10.22 Source side Sn-X-L_TT function.

Interfaces

Sn–X–L_TT_So.

inputs	outputs
at Sn–X–L_So_AP Sn–X–L_AI_D Sn–X–L_AI_CK Sn–X–L_AI_FS	**at Sn–X–L_So_AP** Sn–X–L_AI_X_{AT}
at Sn–X–L_So_CP Sn–X–L_CI_X_{AT}	**at Sn–X–L_So_CP** Sn–X–L_CI_D Sn–X–L_CI_CK Sn–X–L_CI_FS

Processes

The process definitions for this function are the same as for the Sn–X_TT_So function described in Section 10.1.3 with the following technology-specific particularization:

- The number of active members X_{AT} in the VCG received from the Sn–Xv/Sn–X–L_A_So function is transferred back to the Sn–X–L/client_A_So function for transparency.

Defects

None.

Consequent actions

None.

Defect correlation

None.

Performance monitoring

None.

Sink side: Sn–X–L_TT_Sk

Symbol

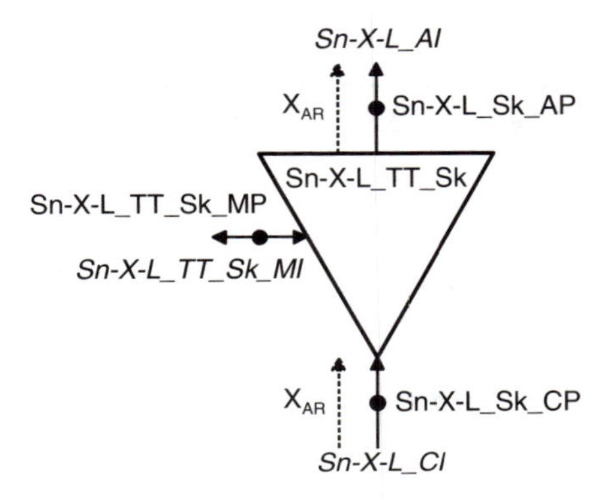

Figure 10.23 Sink side Sn-X-L_TT function.

Interfaces

Sn–X–L_TT_Sk.

inputs	outputs
at Sn–X–L_Sk_CP Sn–X–L_CI_D Sn–X–L_CI_CK Sn–X–L_CI_FS Sn–X–L_CI_SSF Sn–X–L_CI_X_{AR}	**at Sn–X–L_Sk_AP** Sn–X–L_AI_D Sn–X–L_AI_CK Sn–X–L_AI_FS Sn–X–L_AI_TSF Sn–X–L_AI_X_{AR}
at Sn–X–L_TT_Sk_MP Sn–X–L_TT_Sk_MI_SSF_Reported	**at Sn–X–L_TT_Sk_MP** Sn–X–L_TT_Sk_MI_cSSF

Processes

The process definitions for this function are the same as for the Sn–X_TT_Sk function described in Section 10.1.3 with the following technology-specific particularizations:

- The number of active members X_{AR} in the VCG received from the Sn–Xv/Sn–X–L_A_Sk function is transferred to the Sn–X–L/client_A_Sk function.

Defects

None.

Consequent actions

$$aTSF \leftarrow CI\_SSF$$

Defect correlation

$$cSSF \leftarrow CI\_SSF \text{ and } SSF\_Reported$$

Performance monitoring

None.

10.2.4 Sn trail termination function

See Section 10.1.3 for the description. This function does not have any LCAS specific functionality. However, because LCAS is per definition asymmetrical, the member trails are uni-directional and hence RDI and REI are not supported when LCAS is enabled.

10.2.5 Sn–X–L to client signal adaptation function

This function requires client specific inputs, outputs and processes. It is included here to show that the X_{AT} and X_{AR} signals received from the VCG adaptation function have to be processed by this function to adjust the client signal to the available bandwidth. Figure 10.24 shows the bi-directional model of this adaptation function.

Bi-directional model: Sn–X–L/<client>_A

Symbol

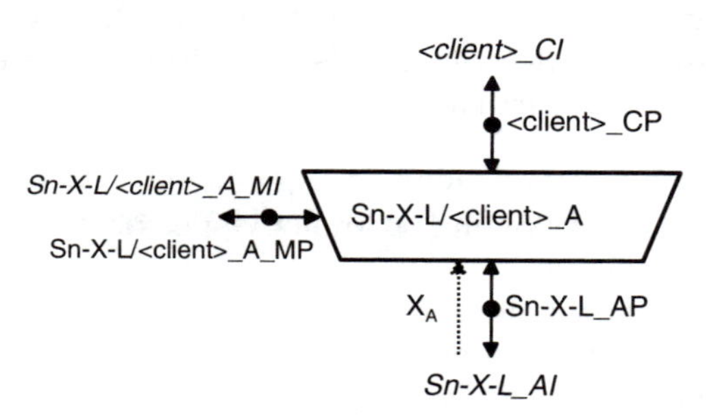

Figure 10.24 Bi-directional Sn-X-L/<client>_A function.

This model can be further decomposed into the uni-directional models of the source and sink side (see Figures 10.25 and 10.26).

Source side: Sn–X–L/<client>_A_So

Symbol

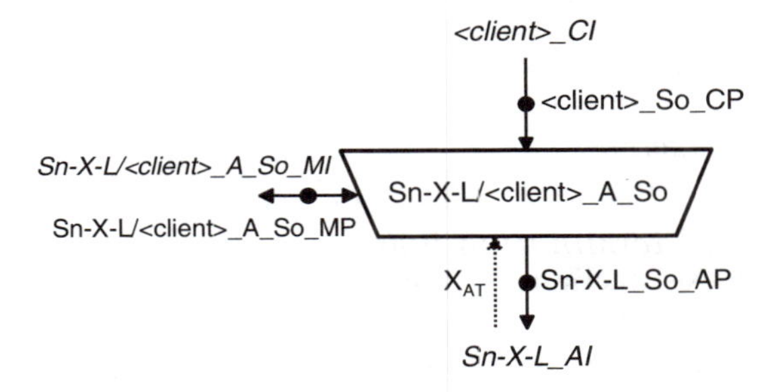

Figure 10.25 Source side Sn-X-L/<client>_A function.

Interfaces

Sn–X–L/<client>_A_So.

inputs	outputs
at Sn–X–L_So_AP Sn–X–L_X_{AT}	**at Sn–X–L_So_AP** Sn–X–L_D Sn–X–L_CK Sn–X–L_FS
at <client>_So_CP <client>_CI_nn	**at <client >_So_CP** client>_CI_nn
at Sn–X–L/ <client>_A_So_MP Sn–X–L/<client>_A_So_MI_nn	**at Sn–X–L/ <client> _A_So_MP** Sn–X–L/<client>_A_So_MI_nn

Processes

The process has the following technology-specific particularizations:

- The incoming client signal is mapped into the payload area of the Sn–X–L signal. The number of active members X_{AT} determines the available payload size. X_{AT} is received from the Sn–Xv/Sn–X–L_A_So function via the VCG Sn–X–L_TT_So function.

Defects

None.

Consequent actions

None.

Defect correlation

None.

Performance monitoring

None.

Sink side: Sn–X–L/<client>_A_Sk

Symbol

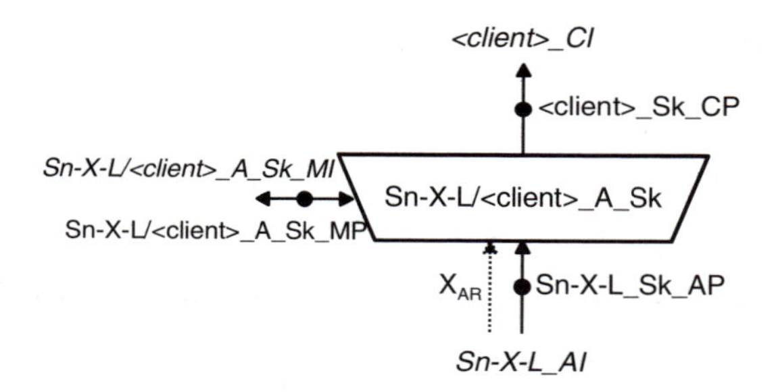

Figure 10.26 Sink side Sn-X-L/<client>_A function.

Interfaces

Sn–X–L/<client>_A_Sk.

inputs	outputs
at Sn–X–L_Sk_AP Sn–X–L_D Sn–X–L_CK Sn–X–L_FS Sn–X–L_TSF Sn–X–L_X_{AR}	
at < client >_Sk_CP <client>_CI_nn	**at < client >_Sk_CP** <client>_CI_nn

at Sn–X–L/< client >_A_Sk_MP	at Sn–X–L/ < client > _A_Sk_MP
Sn–X–L/<client>_A_Sk_MI_Active Sn–X–L/<client>_A_Sk_MI_nn	Sn–X–L/<client>_A_Sk_MI_cPLM Sn–X–L/<client>_A_Sk_MI_AcSL Sn–X–L/<client>_A_Sk_MI_nn

Processes

The process has the following technology-specific particularizations:

- The outgoing client signal is retrieved from the payload area of the Sn–X–L signal. The number of active members X_{AT} determines the actual payload size. X_{AT} is received from the VCG function Sn–Xv/Sn–X–L_A_Sk transferred by the Sn–X–L_TT_Sk function.

VC–n specific sink process

C2: The signal label is recovered from the C2 byte as per ITU-T Rec. G.806 (2004) clause 6.2.4.2. The signal label value '<client> layer mapping' shall be expected. The accepted value of the signal label AcSL is also available at the Sn–X–L/ETH_A_Sk_MP.

Defects

dPLM and <client> layer specific defects.

Consequent actions

<client> layer specific consequent actions.

Defect correlation

cPLM and <client> layer specific probable causes.

Performance monitoring

<client> layer specific.

10.3 VCAT NETWORK MODEL

Figure 10.27 depicts an example functional model of a virtual concatenation sub-layer network illustrating the transport of a single contiguous client signal <client>_CI in a group of X individual member signals Sn_CI. Whether LCAS is enabled or disabled is irrelevant for this model. For simplicity only two member network connections are shown. Diverse routing is illustrated by the different sub-network connections in the Sn layer network for the intermediate nodes. The compound functions 'srv' represent the adaptation and termination of the server layer that is used to support the X individual member network connections Sn_NC[i].

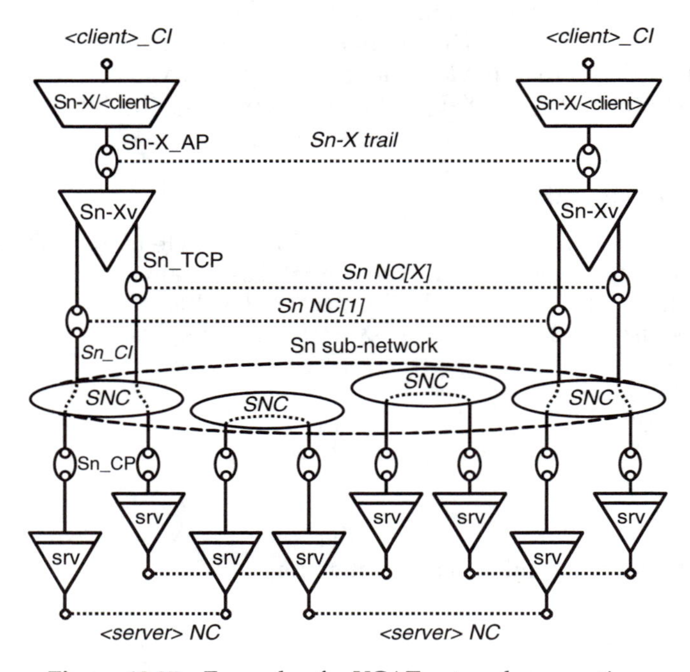

Figure 10.27 Example of a VCAT network connection.

10.4 S4–Xc TO S4–Xc INTERWORKING FUNCTION

Contiguous concatenated characteristic information Sn–Xc_CI and virtual concatenated characteristic information Sn–Xv_CI do have exactly the same payload capacity if they have the same index X. The adapted information of both signals is the same: Sn–X_AI. This feature enabled the deployment of an application specific compound function—the Sn–Xc to Sn–Xv interworking function—a feature that provides a new possibility for layer network interworking. A special symbol has been defined for this function that provides the transformation between contiguous concatenation and virtual concatenation: the interworking function. The application is limited to concatenated S4 signals; more specifically X is limited to the values 4, 16, 64 and 256. The symbol that has been defined for the interworking function is depicted in Figure 10.28.

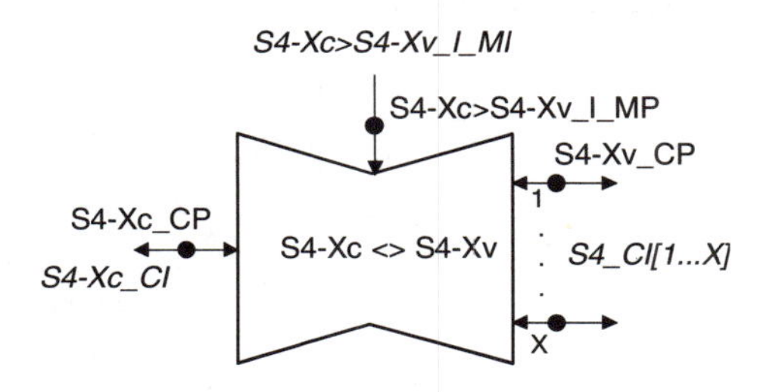

Figure 10.28 S4-XC<>S4-Xv compound function.

This function can be decomposed into the following atomic functions: the virtual concatenation trail termination function Sn–Xv_TT and the contiguous concatenation trail termination function Sn–Xc_TT. Figure 10.29 illustrates the decomposition. For a description of the

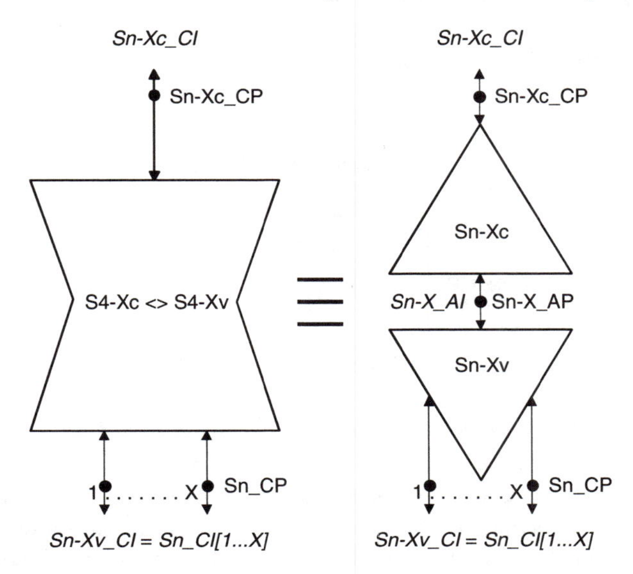

Figure 10.29 Decomposition of the Sn-Xv<>Sn-Xc function.

Sn–Xv_TT function, see Section 10.1. The description of the Sn–Xc_TT function is similar to that of the Sn_TT function described in Section 10.1.4 because the only difference is the size of the payload area of the terminated signal, i.e. it is X times the size of an Sn structure.

The following sections describe the functionality of the interworking function starting with the bi-directional model in Figure 10.30.

Bi-directional model: Sn–Xc <> Sn–Xv

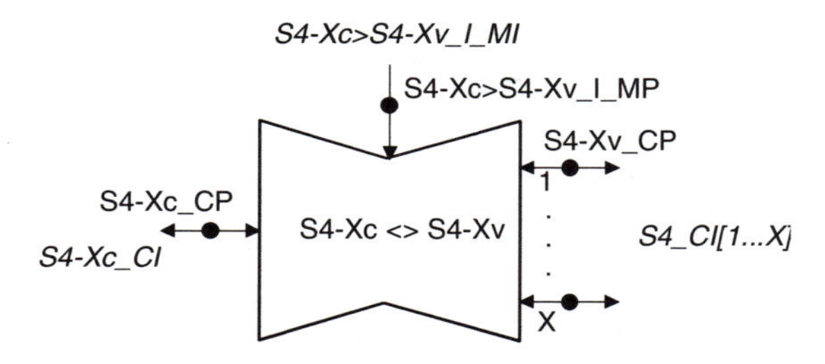

Figure 10.30 Bi-directional S4-Xc<>S4-Xv model.

This model can be further decomposed into the contiguous-to-virtual interworking function (Figure 10.31), and the virtual-to-contiguous interworking function (Figure 10.33).

Contiguous to virtual: S4–Xc > S4–Xv_I

Symbol

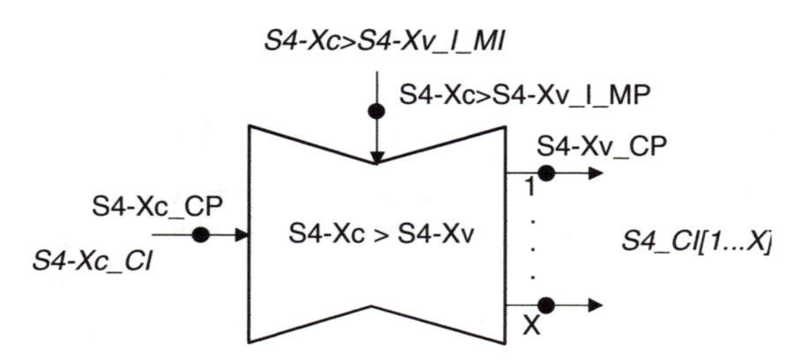

Figure 10.31 S4-Xc>S4-Xv interworking function.

This uni-directional interworking function will convert a contiguous concatenated signal S4–Xc_CI into a virtual concatenated signal S4–Xv_CI that is equivalent to X individual signals S4_CI. Currently, values of X = 4, 16, 64 and 256 are allowed.

Interfaces

S4–Xc>S4–Xv_I.

inputs	outputs
at S4–Xc_CP S4–Xc_CI_D S4–Xc_CI_CK S4–Xc_CI_FS S4–Xc_CI_SSF **at S4–Xc>S4–Xv_I_MI** S4–Xc>S4–Xv_I_MI_TIEn S4–Xc>S4–Xv_I_MI_TxTI[2. . . X]	**at S4–Xv_CP** S4_CI[1. . . X]_D S4_CI[1. . . X]_CK S4_CI[1. . . X]_FS S4_CI[1. . . X]_SSF

Processes

- The received TTI in the S4–Xc overhead will either be sent in the overhead of all S4s of the S4–Xv or only in the overhead of the first S4 of the S4–Xv. In the latter case (if provisioned TIEn = true) the TTI of the remaining S4s has to be provisioned by the NMS.
- The number of bit errors detected in the S4–4c signal will be added to the BIP calculated before insertion in the first S4 (SQ = 0) of the S4–4v. For the remaining S4s of the S4–Xv the BIP is calculated and inserted without change.
- The signal label of the S4–Xc will be inserted in the overhead of all individual S4s of the S4–Xv signal.
- The REI of the S4–Xc will be inserted in the first S4 of the S4–Xv. The REI of all other S4s of the S4–Xv will be set to 0.
- The RDI of the S4–Xc will be inserted in all S4s of the S4–Xv.
- If the Tandem Connection (TC) Overhead (TCOH) bits 1 to 4, the Incoming Error Count (IEC), of the S4–Xc contain the code '1110' (Incoming AIS), the TCOH bits 1 to 4 of all S4s of the S4-v will be set to '1110'. If the IEC of the S4–Xc contains the code '0000' (part of TC Unequipped), the IEC of all S4s of the S4–Xv will be set to '0000'. Otherwise, the IEC of the S4–Xc will be copied to the IEC of the first S4 of the S4–Xv and the IEC of all other S4s of the S4–Xv shall be set to an '1001' indicating IEC = 0. The TC Overhead bits 5 to 8 will be inserted in the first S4 of the S4–Xv. The TCOH bits 5 to 8 of all other S4s of the S4–Xv will be set to 0.

- The remaining overhead bytes of the S4–Xc will be inserted in the first S4 of the S4–Xv signal. The remaining overhead bytes of all other S4s of the S4–Xv will be set to 00h.

The processes described above and their relations are depicted in Figure 10.32.

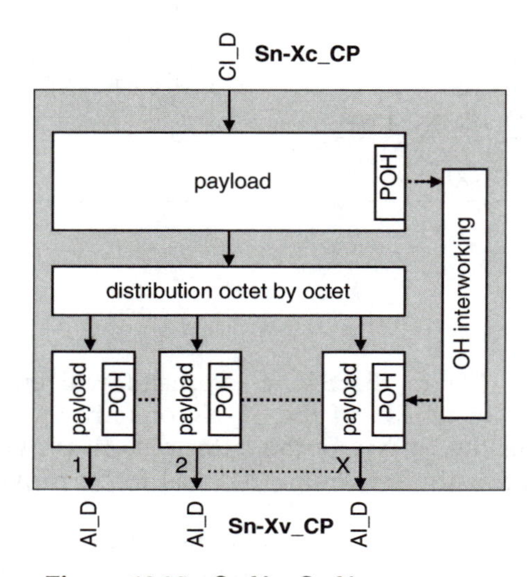

Figure 10.32 Sn-Xc>Sn-Xv processes.

Defects

None.

Consequent actions

A received SSF will be sent as SSF on all S4s of the S4–Xv and the signal of all S4s is replaced by AIS.

$$aAIS \leftarrow CI\_SSF$$
$$aSSF[n] \leftarrow CI\_SSF$$

Defect correlation

None.

Performance monitoring

None.

Virtual to contiguous: S4–Xv>S4–Xc_I

Symbol

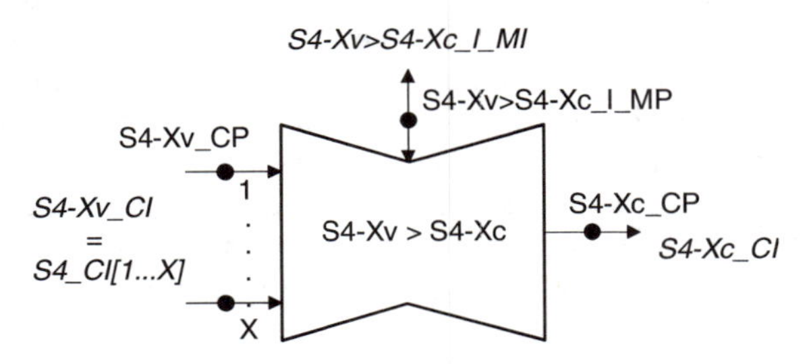

Figure 10.33 Sink side S4-Xv>S4-Xc function.

This uni-directional interworking function will convert a virtual concatenated signal S4–Xv_CI transported by X individual signals S4_CI into a contiguous concatenated signal S4–Xc_CI. Currently, values of X = 4, 16, 64 and 256 are allowed.

Interfaces

S4–Xv>S4–Xc_I.

inputs	outputs
at S4–Xv_CP S4_CI[1... X]_D S4_CI[1... X]_Ck S4_CI[1... X]_FS S4_CI[1... X]_SSF	**at S4–Xc_CP** S4–Xc_CI_D S4–Xc_CI_CK S4–Xc_CI_FS S4–Xc_CI_SSF
at S4–Xc>S4–Xv_I_MI S4–Xv>S4–Xc_I_MI_Tpmode S4–Xv>S4–Xc_I_MI_SSF_Reported S4–Xv>S4–Xc_I_MI_ExTI[1... X] S4–Xv>S4–Xc_I_1second S4–Xv>S4–Xc_I_TIMdis[1... X]	**at S4–Xc>S4–Xv_I_MI** S4–Xv>S4–Xc_I_MI_cTIM[1... X] S4–Xv>S4–Xc_I_MI_cUNEQ[1... X] S4–Xv>S4–Xc_I_MI_cSSF[1... X] S4–Xv>S4–Xc_I_MI_AcTI[1... X] S4–Xv>S4–Xc_I_MI_cLOM[1... X] S4–Xv>S4–Xc_I_MI_cSQM[1... X] S4–Xv>S4–Xc_I_MI_cLOA S4–Xv>S4–Xc_I_MI_AcSQ[1... X]

Processes

The following processes are performed per individual S4 in the S4–Xv:

- The TTI is recovered from S4 overhead. The accepted value is reported as AcTI and it is compared to the provisioned expected (ExTI) value. A detected mismatch defect (dTIM) is reported (cTIM) if the reporting is not disabled (TIMdis). (Note: If no individual TTIs are configured for the S4s in the S4–Xc>S4–Xv_I function, the expected TTI for all S4s has to be set equal to the expected TTI of the first S4, or TPmode has to be set for these S4[2. . .(X − 1)].)
- The signal label byte C2 will be recovered. It will be used to detect (dUNEQ) and report (cUNEQ) the unequipped defect or the AIS condition (dAIS).
- The SQ will be recovered and after acceptation it will be reported as AcSQ. A new sequence number is accepted if it has m times the same value with $3 \leq m \leq 10$.

The following functions will be performed for the S4–Xc:

- The H4 byte of the S4–Xc will be set to '0'.
- The received TTI of the first S4 of the S4–Xv shall be inserted to the S4–Xc.
- The BIP–8 will be calculated for each S4 frame of the S4–Xv and compared with the related received BIP–8 to determine the bit errors per S4. The bit errors of all S4s of the S4–Xc will be added together and the result will be added to the BIP–8 calculated for S4–Xc frame. The resulting BIP–8 will be inserted in the S4–Xc overhead.
- The signal label of the first S4 of the S4–Xv shall be inserted in the S4–Xc.
- The REI values of all S4s of the S4–Xv will be added together. The result will be inserted in the S4–Xc.
- If the RDI of any S4 of the S4–Xv is set, the RDI of the S4–Xc shall be set to '1'.
- If the TCOH bits 1 to 4 (IEC) of any S4 of the S4–Xv contain the code '1110' (Incoming AIS), TCOH bits 1 to 4 of the S4–Xc shall be set to '1110'. If the IEC of the first S4 of the S4–Xv contain the code '0000' (TC Unequipped), the IEC of the S4–Xc shall be set to '0000'. Otherwise, the IEC values of all S4s of the S4–Xv will be added together and inserted as IEC in the S4–Xc. TCOH bits 5 to 8 of the first S4 of the S4–Xv will be copied to TCOH bits 5 to 8 of the S4–Xc.

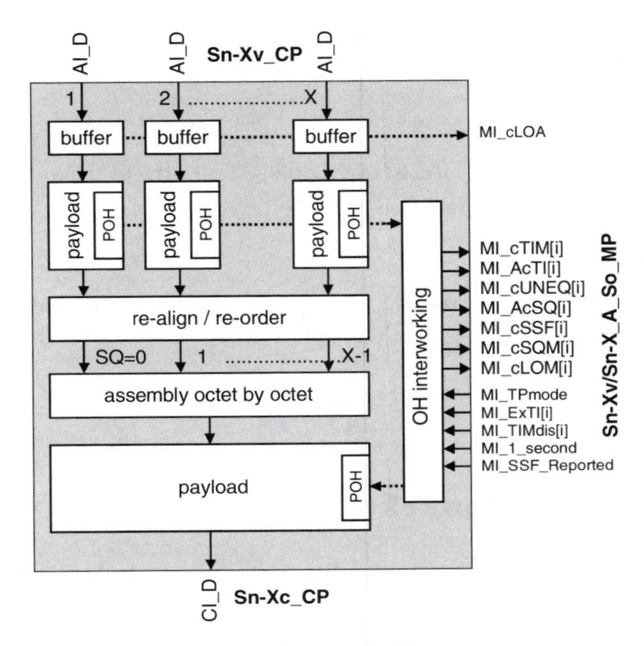

Figure 10.34 Sn-Xv>Sn-Xc processes.

The remaining overhead bytes of the first S4 of the S4–Xv will be copied to the S4–Xc. The processes described above and their relations are depicted in Figure 10.34.

Defects

For each individual VC–4, the defects dUNEQ, dAIS and dTIM shall be detected as described above.

After re-aligning by using the multi-frame process, and re-ordering by using the re-sequence process, the X individual VC–4s the following defects are detected for the VCG:

- *Loss Of Multi-frame* (dLOM): The VCAT overhead multi-frame cannot be recovered from the received signal.
- *Sequence number Mismatch* (dSQM): The accepted SQ number (AcSQ) of a member does not match the expected SQ number (EcSQ). The expected SQ number of member p in a VCG is (p − 1).
- *Loss of Alignment* (dLOA): The differential delay, i.e. the difference in time between the member experiencing the least propagation

delay and the member experiencing the most propagation delay in the network, exceeds the implemented alignment buffer.

Consequent actions

If one of the defects dTIM, dLOM, dSQM or dLOA is detected, the S4–Xc signal will be replaced by an AIS signal.

Defect correlation

Based on the detected defects this process will determine the most probable causes: cUNEQ, cTIM, cSSF, cLOM, cSQM and cLOA.

Performance monitoring

None.

10.5 VCAT-CCAT INTERWORKING NETWORK MODEL

The Sn–Xc <> Sn–Xv compound function defined in ITU-T Re. G.805 (2000) is located between an Sn–Xc and an Sn–Xv layer network with

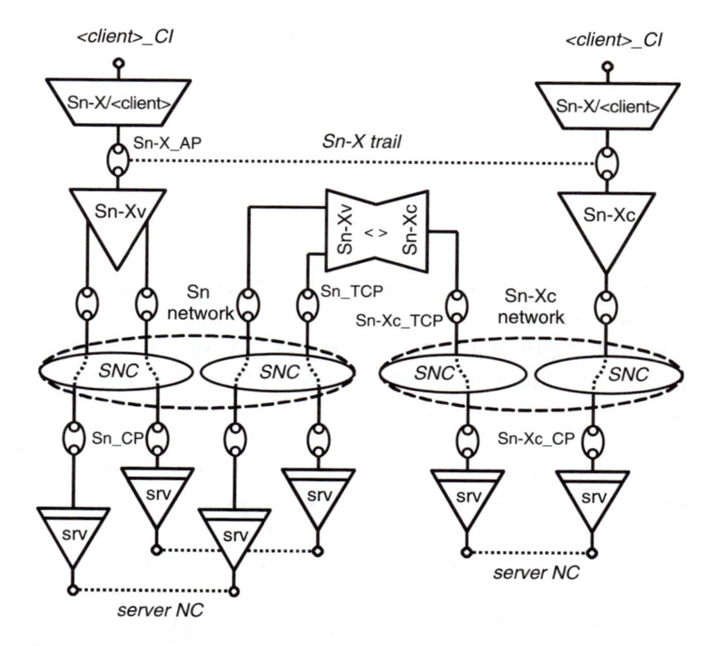

Figure 10.35 Example of VCAT CCAT network interworking.

the same index X, where the latter is in fact an Sn layer network with X separate Sn sub-network connections. Figure 10.35 shows an example functional network architecture illustrating the transport of a contiguous client signal <client>_CI over a network, partly as a virtual concatenated signal and partly as a contiguous concatenated signal. The interworking processing function performs the semantically 'transparent' conversion of the virtual concatenation trail overhead to contiguous concatenation trail overhead and vice versa. An Sn–X trail may contain one or more interworking processing functions. The compound functions 'srv' represent the adaptation and termination of the server layer that is used to support the transport of the X individual member signals Sn_CI[i] or the single contiguous concatenated signal Sn–Xc_CI.

11

Example functional models to exercise

In this chapter a number of examples are provided to illustrate the use of functional modeling at different levels in the network. They include the model used by a device implementer, a model used by a unit level designer, a network element description, and a network application.

A functional model can be read best if it is drawn on a single page or sheet. This can be achieved, without losing the required overall functionality, by using compound functions, and/or by removing sublayers, e.g. protection sublayers or performance monitoring sublayers, and providing separated figures containing the detailed sublayer information. The compound functions can be expanded to explain or describe certain detailed aspects of a functional model in separate figures.

11.1 DEVICE LEVEL FUNCTIONAL MODEL

In the first example, the most detailed functional model is provided containing all possible mappings of a 2.048 kbit/s PDH signal into a VC–12 container. Also, it includes a suggestion for a compound function that can be used in less detailed functional models of a unit level or system level description.

Figure 11.1 shows a very detailed model of the functions that are present between the physical 2.048 kbit/s interface signal and the VC–12 signal. A detailed description of the atomic functions used in this figure is given in ITU-T Rec. G.705 (2000).

SDH/SONET Explained in Functional Models Huub van Helvoort

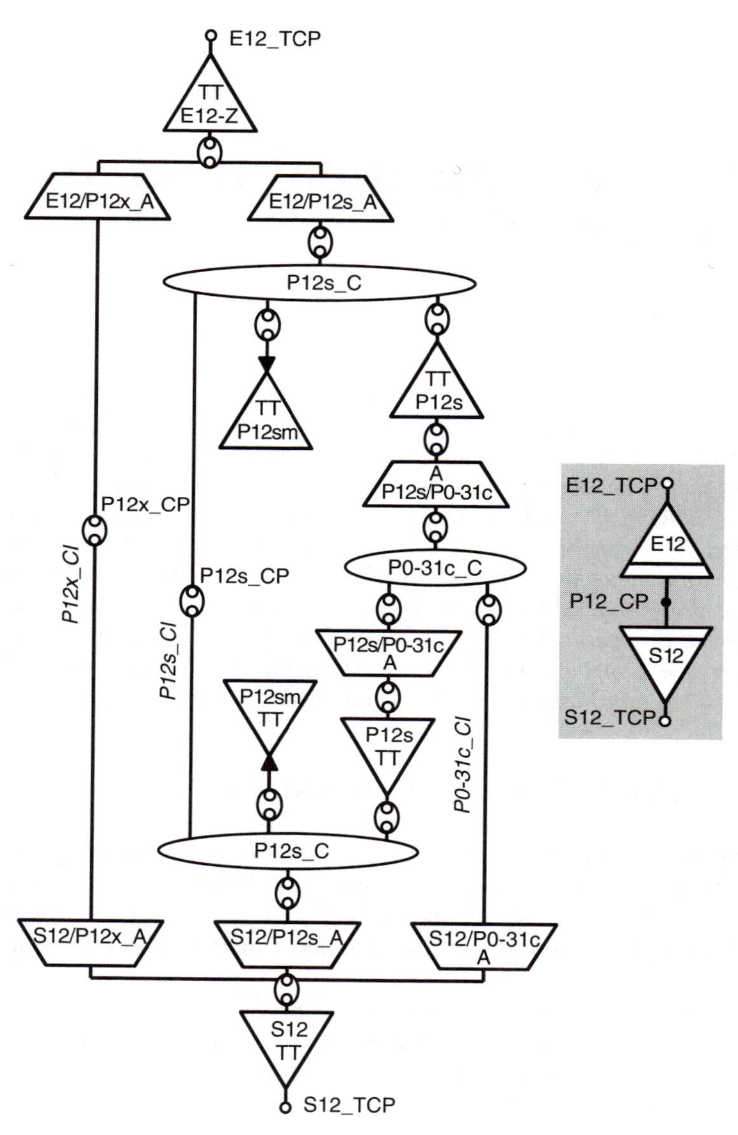

Figure 11.1 Detailed 2 Mbit/s model and compound representation.

The 2.048 kbit/s PDH or E12 characteristic information present at the connection point E12_TCP is described in ITU-T Rec. G.703 (1998) clause 9. It is terminated in the E12-Z_TT function where Z indicates the supported cable impedance, 75 Ω or 120 Ω. Now, the recovered 2.048 kbit/s signal can either be transported as a constant bitrate signal (disregarding any possible framing) or as an ITU-T Rec. G.704 (1998) or G.706 (1991) framed signal. This is shown by the presence of two adaptation functions: E12/P12x_A and E12/P12s_A. The P12x_CI represents the CBR signal and P12s_CI represents the 32 octet based framed signal. The S12/P12x_A provides a-synchronous (bit) mapping of the CBR signal in a VC–12 according to ITU-T Rec. G.707 (2003) clause 10.1.4.1. The S12/P12s_A provides byte synchronous (octet) mapping according to ITU-T Rec. G.707 (2003) clause 10.1.4.2.

A 2.048 kbit/s link may be partitioned into sections to improve fault localization and performance monitoring. In this case the framed 2.048 kbit/s signal can be terminated in the P12s_TT function. The remaining P0-31c_CI signal carries the 31 concatenated 64 kbit/s payload octets of the E12 signal at a total rate of 1984 kbit/s. This signal can be mapped into a VC–12 according to ITU-T Rec. G.707 (2003) clause 10.1.4.3 in the S12/P0-31c_A function while using the performance information provided by the associated VC–12 for the section carried through the SDH network. Also, it is possible to start immediately a new E12 section by adding again framing information to the P0-31c_CI signal in the P12s/P0-31c_A function and use byte synchronous mapping for transport of the signal in a VC–12. The P12sm_TT_Sk monitors can be used to verify the performance of the transported P12x_CI and P12s_CI signals, e.g. as required by a *Service Level Agreement* (SLA).

This is the level of detail used by an implementer of this functionality, e.g. a device designer or a software designer responsible for provisioning.

If this level of detail is not required the model depicted in the shaded area of Figure 11.1 can replace the detailed model. It shows two compound functions each existing of at least a termination function and an adaptation function. The compound functions can be used in a model to describe (parts of) equipment.

11.2 EQUIPMENT DETAILED FUNCTIONAL MODEL

In Chapter 3, Section 3.2, a description is provided of partitioning based on the physical properties of a network and the capabilities of its

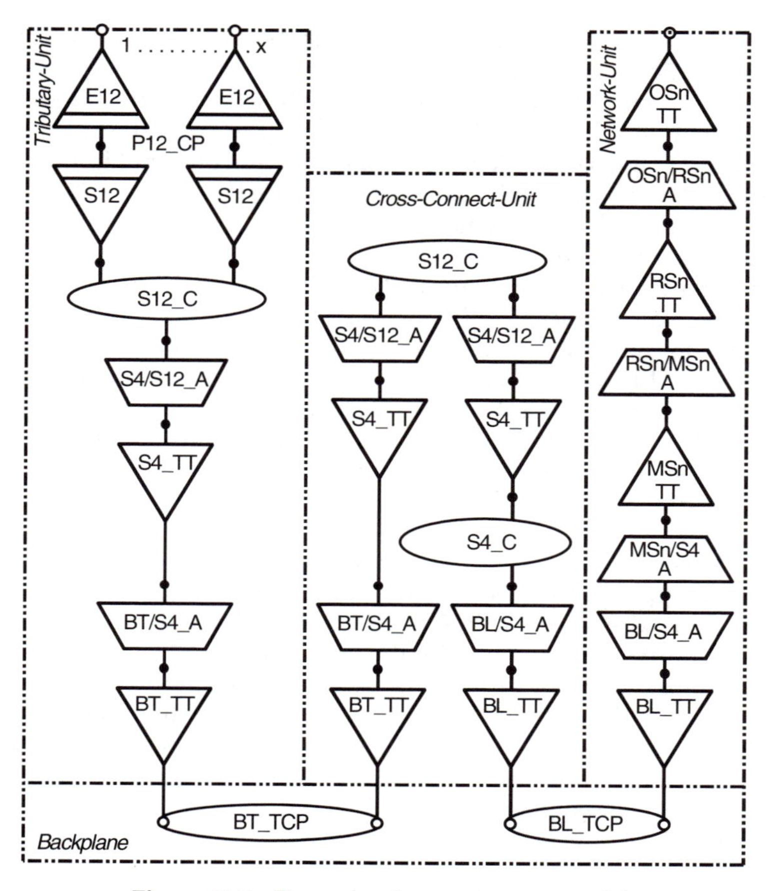

Figure 11.2 Example of an equipment model.

network elements. Partitioning can also be forced by the physical properties of the equipment itself. Figure 11.2 shows an example of equipment partitioning. It illustrates how functions are grouped together, forced by the physical constraints. Here, the equipment dimensions and internal and external connectivity determine the partitioning of functions, e.g. the partitioned lower order connection function S12_C or the electrical properties of the backplane. The expansion methodology described in Chapter 4, Section 4.2.3 has been used to provide the internal, i.e. backplane, connectivity.

Depicted are three units that may exist in an add-drop multiplexer (ADM) and the backplane:

- a tributary unit that supports x individual 2.048 kbit/s PDH tributaries (E12) and a lower order (sub-network) connection function, providing, in this example, the lower order S12 connectivity;
- a cross-connect unit that supports the lower order and higher order connectivity required to provide the add/drop capability and that also takes care of connecting the through traffic;
- a network unit that supports an SDH interface for connecting the ADM to the network;
- a backplane is used to interconnect all units and uses proprietary, implementation specific, backplane signals for tributary units BT_CI and network units BL_CI. These backplane signals require their own specific adaptation and termination functions as shown in Figure 11.2 by the BT/S4_A, BT_TT, BL/S4_A and BL_TT symbols.

A user who is interested only in the layer network topology will represent the VC–12 layer in this example network element as depicted in Figure 11.3. In this abstract model one can identify the

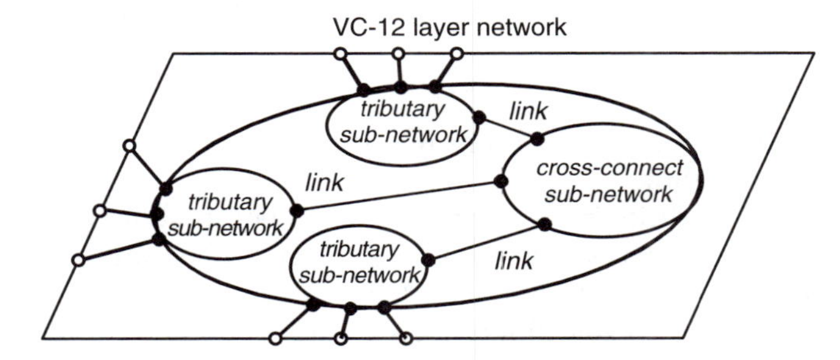

Figure 11.3 Abstract model representing a network element VC–12 layer.

tributary sub-networks with the ports representing the external interfaces, the cross-connect sub-network without external ports and the connecting links, i.e. the backplane connections. In this example, the backplane connection has a VC–4 structure hence each link has a capacity to transport up to 63 VC–12s. A developer of the network element connection management application can use this abstract

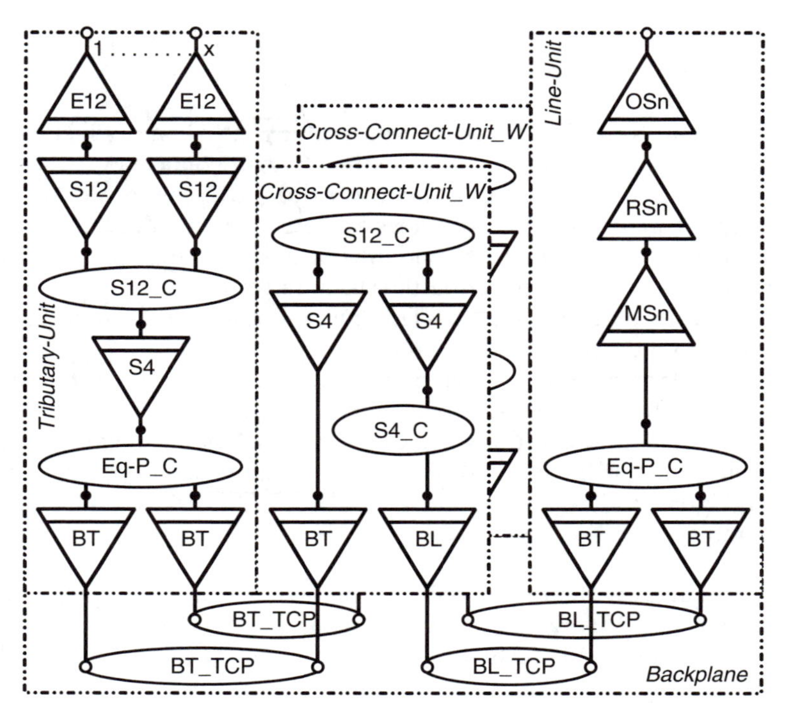

Figure 11.4 Example 1 of an equipment model with equipment protection.

model as a reference and will use the functional model to retrieve the detailed information.

Even more partitioning is required when equipment protection is used to improve the availability of the system. In the example given in Figure 11.4, equipment protection has been provided by both doubling the backplane signals and doubling a complete unit.

In this case the cross-connect unit is duplicated, i.e. a working unit and a protection unit are present in the system. This is similar to the methodology used for trail protection explained in Chapter 9, Section 9.2.1. The connection functions Eq-P_C provide the protection switching capability. The fact that an ADM requires two identical line interfaces can also be used in the partitioning of the system to provide equipment protection. In this case the cross-connect unit needs to support the MS trail protection methodology as described in Chapter 9, Section 9.2.1.2. This is illustrated in Figure 11.5.

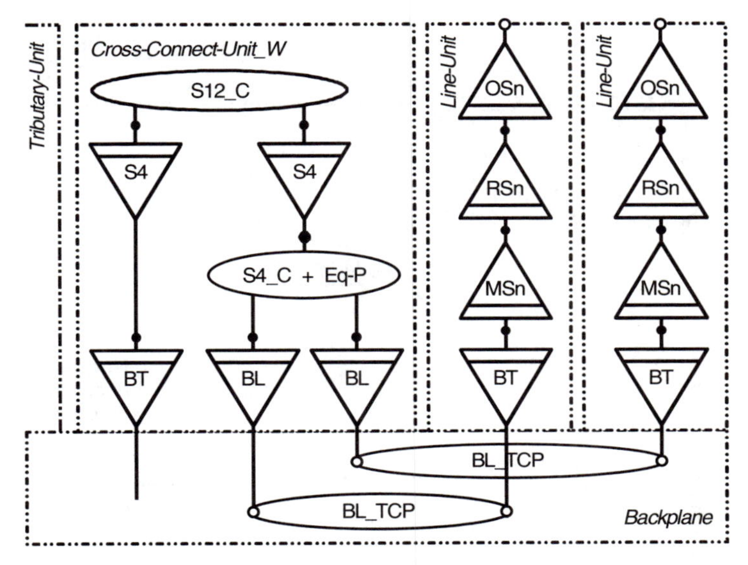

Figure 11.5 Example 2 of an equipment model with equipment protection.

Although the functional models from Figures 11.3, 11.4 and 11.5 can be used to show the actual data flow through the equipment, they are developed for use by a unit designer and therefore may contain too much information for a user who is interested only in a model showing the functional flow of information through the network element. Figure 11.6 shows the same information flow as depicted in Figures 11.3, 11.4 and 11.5, however, all partitioning information related to particular implementation decisions has been removed. This figure also indicates which different layer networks exist in the network element.

11.3 NETWORK ELEMENT FUNCTIONAL MODEL

By removing all the implementation details an equipment vendor can describe network elements without giving away proprietary information. A buyer can use the functional equipment model to verify that a network element can be used in the network for which it is intended.

Figure 11.7 shows an example of a network element to be used as an ADM. The original ADMs had two SDH line interfaces at the network

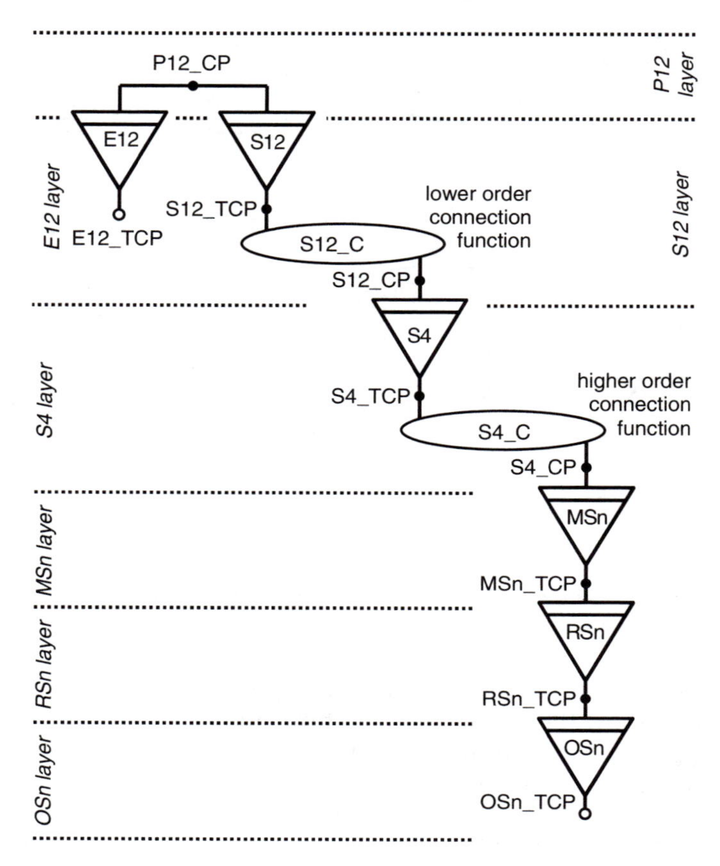

Figure 11.6 Network element simplified functional model.

side and several PDH interfaces at the tributary side and also lower order SDH interfaces. If only one line interface is used or equipped the network element is referred to as a *Terminal Multiplexer* (TM).

The maximum number of interfaces supported per tributary signal may vary as indicated in Figure 11.7. The actual number of tributary interfaces depends on the number of available physical unit slots, the number of interfaces supported per unit and the connection capabilities of the lower order and higher order connection functions. This model does not provide information about the protection capabilities of the ADM. If it is required to show the protection features, the

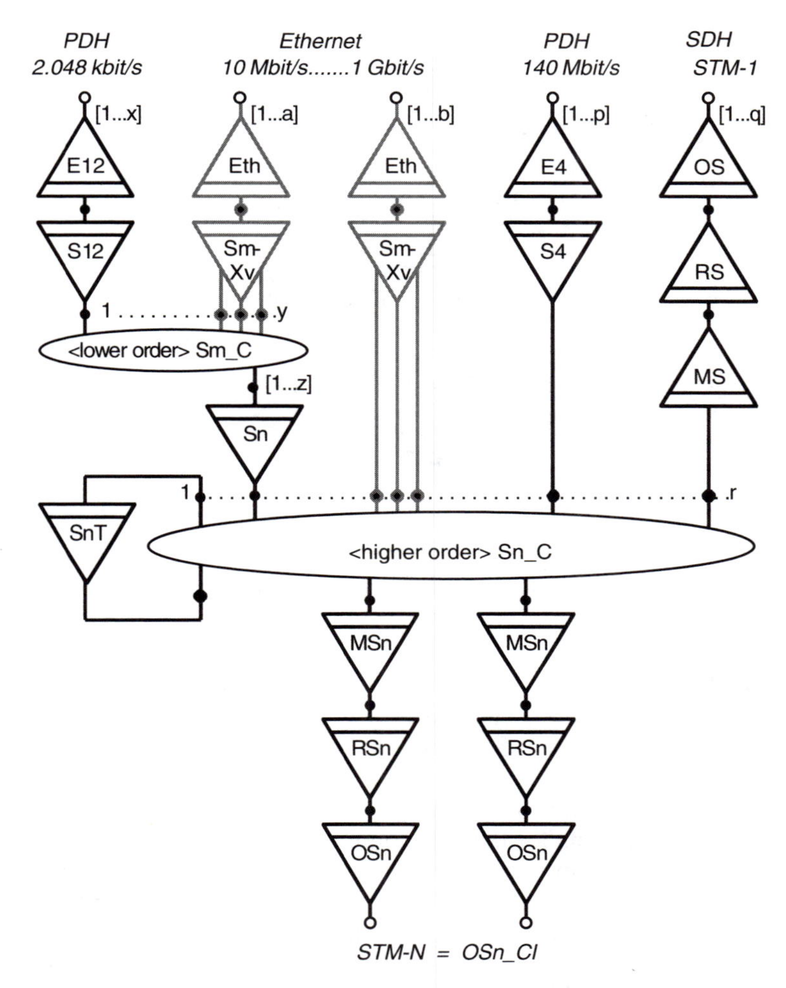

Figure 11.7 Example of an ADM and MSAP.

methodology described in Chapter 9, Section 9.2 (e.g. adding protection layers) can be applied to extend the model.

An additional feature shown in this model is the support of *Tandem Connection Monitoring* (TCM) indicated by the SnT compound function and described in detail in ITU-T Rec. G.707 (2003) Annexes C and D. TCM provides performance monitoring of part(s) of a path, e.g. the part that passes through a different operator domain (see also Chapter 8,

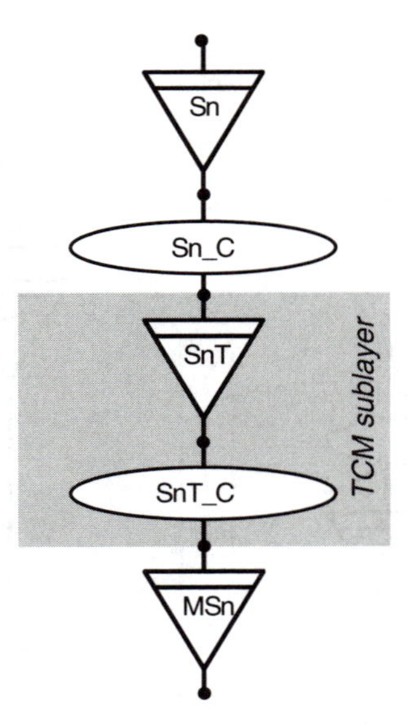

Figure 11.8 TCM sublayer.

Section 8.3.2). When TCM is used an extra overhead sublayer is introduced as illustrated in Figure 11.8. The tandem connection sublayer is an optional sublayer that is located between the multiplex section and path layers. The use of tandem connection is application specific and at the discretion of the operator. It is expected that the principal applications for tandem connection will be in the inter-office network and that tandem connections will generally not be used in applications such as the subscriber access network. Note that only one level of TCM sublayering is allowed in SDH.

The original ADM can be upgraded to become a *Multi Service Transport Platform* (MSTP) by introducing units that support data transport signals, provide interfaces (e.g. Ethernet or FICON) and use GFP and VCAT methodology for the mapping and transport of these signals. These additional features are drawn in gray in Figure 11.7. In this example, the MSP functionality provides only the transport of data signals and can be used for aggregation of these signals in the access part of the network; the network element will be identified as a *Multi Service Access Platform* (MSAP). For the actual switching and routing this application relies on external equipment that can be placed at a strategic or centralized location in the network.

Recently developed network elements serving as an MSTP will have more capabilities to support the transport of data signals. In the functional model provided as an example of an MSTP and shown in Figure 11.9, several different Ethernet interfaces are supported as well as the switching of data. The received PDUs are switched in the Ethernet connection function Eth_C that has bridge or router functionality.

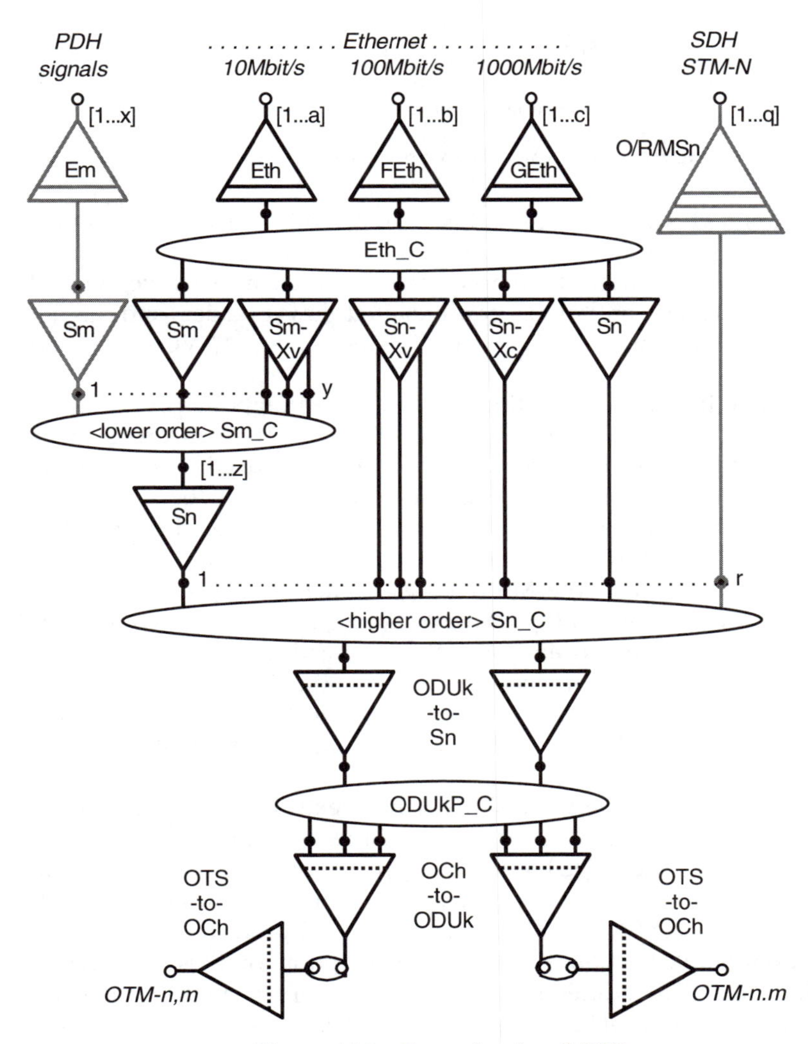

Figure 11.9 Example of an MSSP.

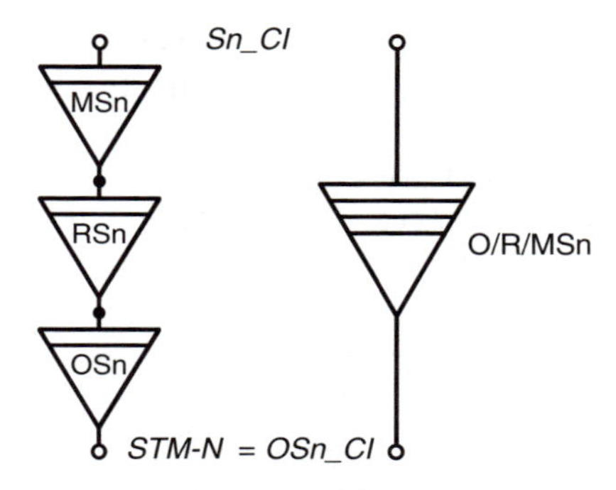

Figure 11.10 Osn-RSn-MSn compound function.

The extended switching capabilities make this network element a *Multi Service Switching Platform* (MSSP). For transport over the SDH network, GFP is used to map packet streams into single, virtual concatenated or contiguous concatenated lower order and higher order SDH containers, i.e. VC–m, VC–m–Xv, VC–n, VC–n–Xv and VC–n–Xc.

Note that because the bindings at the connection points between the MSn layer, the RSn layer and the Osn layer networks are fixed, a compound functional symbol O/R/MSn is introduced. The composition of this compound function is shown in Figure 11.10.

In this function the client signal Sn_CI is first adapted to the SDH *Multiplex Section* (MS) layer by the MSn/Sn_A function, and the MS_TT function adds the MS overhead. Then the signal is adapted to the *Regenerator Section* (RS) layer by the RSn/MSn_A function while the RS_TT function processes the RS overhead. Finally, the signal is adapted to the *Optical Section* (OS) layer by the OSn/RSn_A function where the OS_TT function provides the electrical to optical conversion. The resulting logical characteristic signal OSn_CI is normally referred to as STM-N.

In Figure 10.7, the network side has been extended with *Dense Wave Division Multiplexing* (DWDM) capabilities. In this example, the connection function ODUk_C is used to provide the add/drop functionality of the optical signals present in the connecting fiber. It may also provide SNC/I, SNC/N, and SNC/S protection capabilities (see Chapter 9, Section 9.2.2 for a description of these protection mechanisms). The atomic functions that are represented by the compound functions ODUk–to–Sn, OCh–to–ODUk and OTS–to–OCh are shown in Figures 11.11, 11.12 and 11.13 respectively.

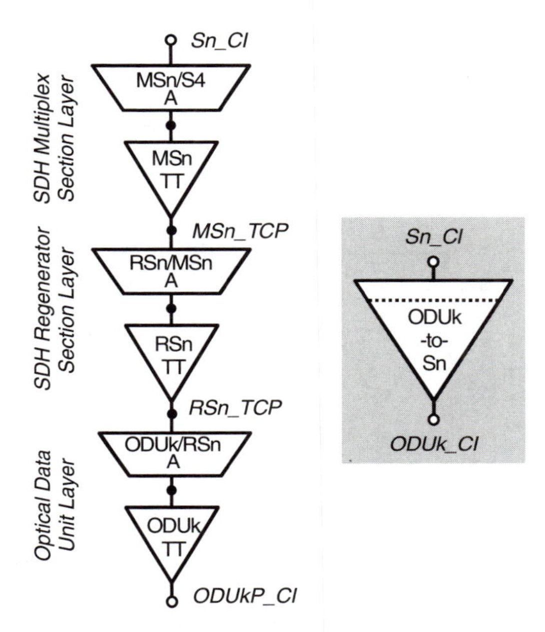

Figure 11.11 The ODUk-to-Sn compound function.

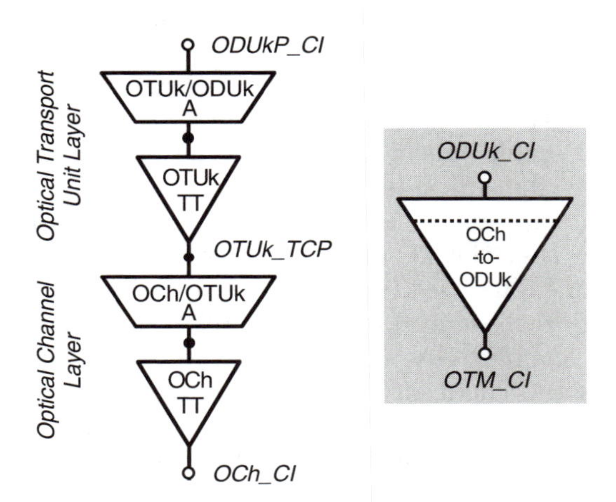

Figure 11.12 The OCh-to-ODUk compound function.

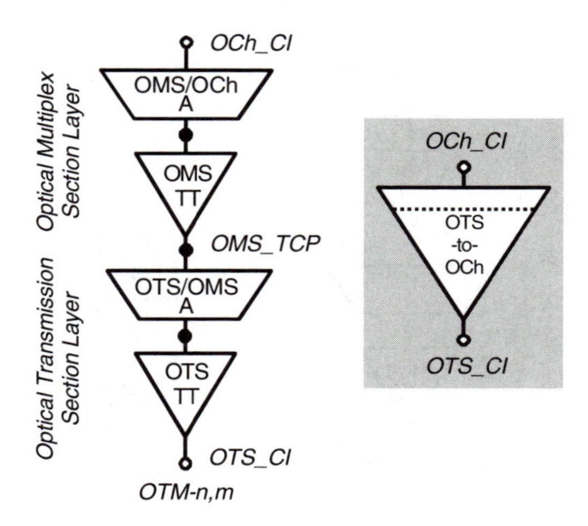

Figure 11.13 OTS-to-OCH compound function.

The STM-N signal, i.e. the original signal on the connecting fibers shown in Figure 11.7, will be treated by the DWDM extension as a client signal that has to be transported on an OTM-n.m server signal in the *Optical Transport Network* (OTN). In the ODUk–to–Sn compound function (see Figure 11.11) the Sn client signal is adapted for transport by an *Optical Channel Data Unit* (ODU) server signal. First, the signal Sn_CI is adapted into an MSn_CI signal and next into an RSn_CI signal as already explained in this section. The next function will adapt this signal for transport in the OTN network, i.e. the ODUkP/RSn-a_A_So function will map the RSn_CI into an OTN container. The structure of the ODUk container facilitates only the mapping of RSn signals with n = 16, 64, 256. The 'a' indicates that the RSn client signal is mapped a-synchronously; bit synchronous mapping is also possible at the source side and the sink side adaptation function ODUkP/RSn_A_Sk supports both mappings (see ITU-T Rec. G.798 (2004) clause 14.3.6 for more details). The ODUkP path overhead is processed by the trail termination function ODUkP_TT. The signal ODUkP_CI at the output is presented to the connection function ODUkP_C, the function that provides the connectivity of the ADM to the OTN. The signal ODUkP_CI is the client signal of the compound function OCh-to-ODUk (see Figure 11.12). The ODUkP client signal is octet synchronous mapped into an *Optical Transmission Unit* (OTUk) by the OTUk/ODUk adaptation function and the OTUk section overhead is processed by the trail termination function OTUk_TT. The resulting signal

OTUk_CI is then adapted by the OCh/OTUk_A function for transport over an *Optical Channel* (OCh) trail. The OCh/OTUk_A function provides the scrambling, forward error correction, and clock recovery processing. After adaptation, the signal is prepared for transfer across the optical fiber by the OCh trail termination function OCh_TT. The characteristic information OCh_CI is the client signal of the compound function OTS–to–OCh (see Figure 11.13) that adapts it for transport in the *Optical Transmission Section* (OTS) layer server signal OTS_CI, also referred to as *Optical Transport Multiplex* OTM–n.m. The OCh client signal is adapted for transport over the *Optical Multiplex Section* (OMS) by the OMS/OCh_A function. The adaptation consists of wavelength assignment and wavelength division multiplexing. The OMS specific overhead is processed by the trail termination function OMS_TT. Next, the OMS signal is adapted into an OTS server trail via the adaptation function OTS/OMS_A. The OTS layer overhead is processed by the trail termination function OTS_TT. This termination function also maps the *Optical Transmission Module* (OTM) signal overhead into the *Optical Supervisory Channel* (OSC). The OTS characteristic information OTS_CI is a combination of the OSC signal and the OTS payload signal and is commonly referred to as OTM–n.m.

11.4 TRAIL CONNECTION MODEL

In this section a description will be provided of how to model a path through a network, starting with the typical topology of an SDH/SONET network, i.e. a number of *Add Drop Multiplexes* (ADM) connected in a loop configuration and normally referred to as a ring. As explained in Chapter 9, Section 9.2.1.2 this topology is very suitable for improving the availability of resources in the network. Figure 11.14 shows a ring configuration with four ADMs.

Also, this figure shows a VC–12 bi-directional trail that transports an E12 signal from node A to node C and vice versa. For this purpose, VC–12 connections have to be made in the switching fabrics of nodes A and C from the tributary side to the line (network) side and in node D VC–12 connections have to be made between the two lines sides as indicated by the dashed lines. The first abstraction is to represent the ring as a VC–12 layer network, with two access points, that contains an S12 sub-network with two ports as depicted in Figure 11.15.

This abstraction is the basis for describing a ring by a functional model. Figure 11.16 shows the model of a VC–12 layer network with

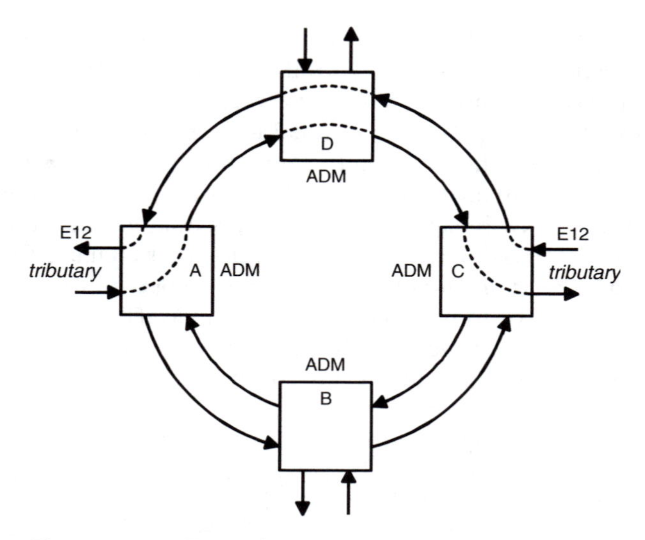

Figure 11.14 Example network ring with four nodes.

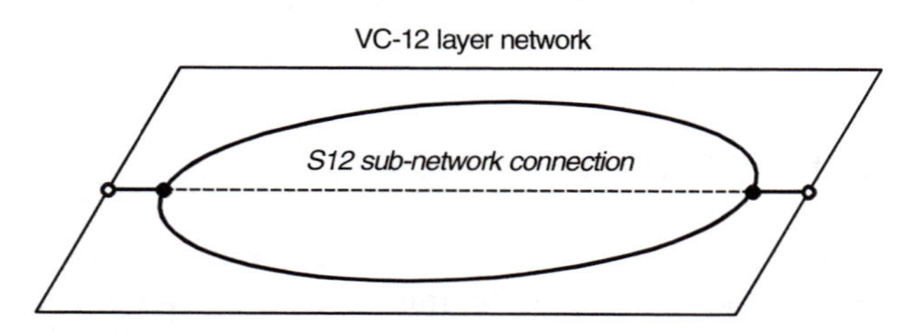

Figure 11.15 Abstract model representing a ring.

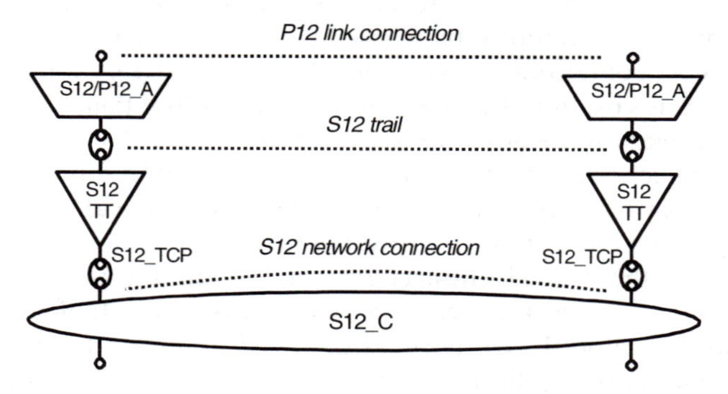

Figure 11.16 Abstract model representing the VC-12 layer network.

all the relevant atomic functions, i.e. the adaptation function S12/P12_A, the trail termination function S12_TT and the connection function S12_C.

The abstract model can be partitioned to show more details, e.g. the actual nodes in the ring. The single sub-network shown in Figure 11.16 is partitioned according to the rules given in Chapter 3, Section 3.2.1 by adding a server layer network, in this case the S4 sub-network. In Figure 11.17 the four S12 sub-networks represent the fabrics of the four ADMs in the ring with the dashed S12 sub-network connections (SNC) representing the VC–12 connections in each node. The *link* connections represent the VC–12 transport capability of the connecting fibers between the individual nodes.

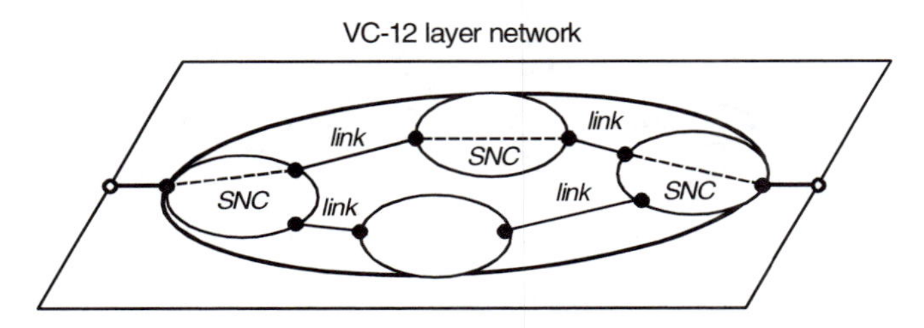

Figure 11.17 Abstract model representing a partitioned ring.

The VC–12 containers are transported over the links by mapping them into a VC–4. If the VC–4 trail used for this purpose in the link between nodes A and D is not the same VC–4 trail used between nodes D and C, these trails are terminated in node D and the VC–12 has to be switched from one trail to the other. This is shown in Figure 11.18.

However, if the VC–4 trail originating in node A is not terminated in node D but switched through in node D and terminated in node C, this figure has to be adapted as depicted in Figure 11.19. Note that any information is cost about the network elements that pass through the VC–4 trail without terminating it.

Returning to the layer network model of Figure 11.17, the fact that a VC–4 trail is passed through in an intermediate network element can only be shown by drawing a direct link between the sub-networks representing the VC–12 connectivity of the individual nodes. The resulting layer network model is provided in Figure 11.20. Here, the VC–4 layer network is added to show the actual VC–4 sub-network connections that are represented by the links in the VC–12 layer network.

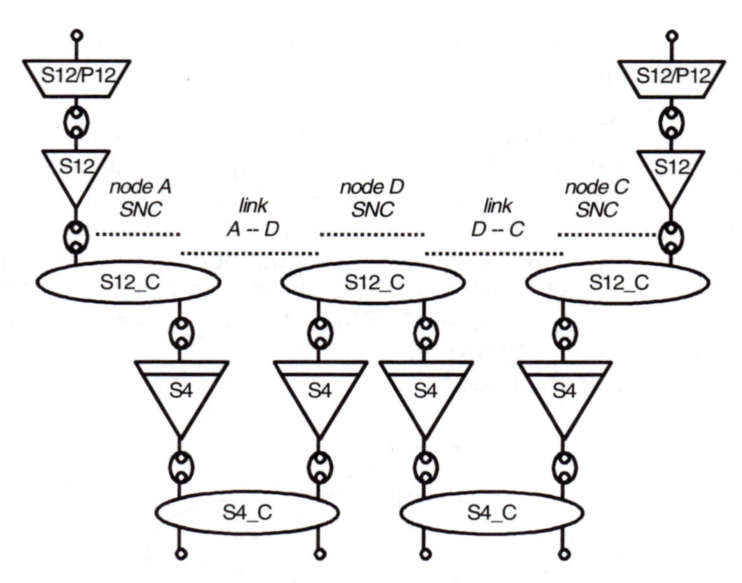

Figure 11.18 Example connection model.

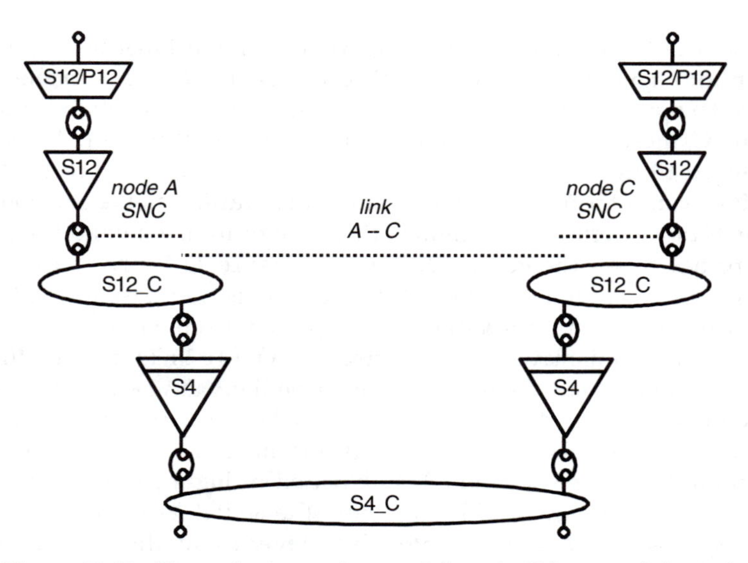

Figure 11.19 Example connection model, node D is passed through.

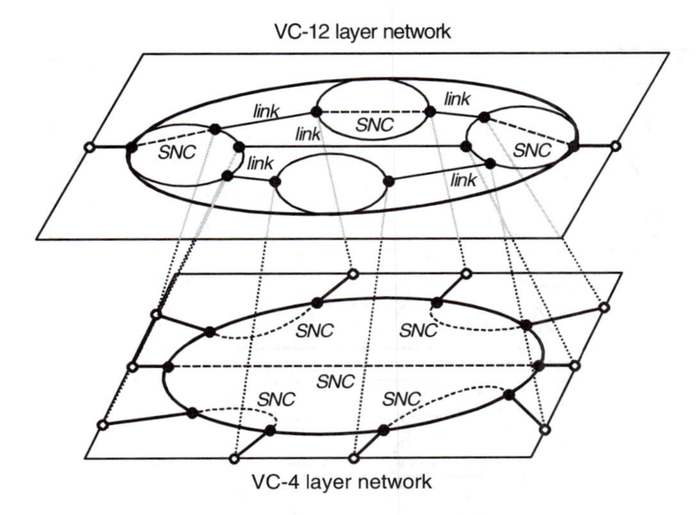

Figure 11.20 Abstract model representing a partitioned and layered ring.

If the user requires that the network topology remains visible in the functional model, then more layers have to be added. In this example, layers down to the optical layer have to be included because the bindings between MSn, Rsn and OSn connection points are fixed and only the OSn sub-network provides the connectivity between adjacent network elements. This is shown in Figure 11.21. The sequence of Msn, RSn and OSn adaptations and terminations is replaced by the same compound function O/R/MSn that is used and described in Section 11.3 (Figure 11.10).

The related abstract model can show the required level of detail as well. This is depicted in Figure 11.22. Note that the two cases mentioned above are shown in this figure:

- case one, where a VC–12 is transported from node A to node C being a member of two different (VC–4) links in the VC–12 layer network requiring a connection in the VC–12 sub-network of node D;
- case two, where a VC–12 is transported from node A to node C being a member of a (VC–4) link between these nodes.

This is visible especially in the VC–12 layer network, i.e. the server links between the sub-networks. Note that the links in the VC–12 layer

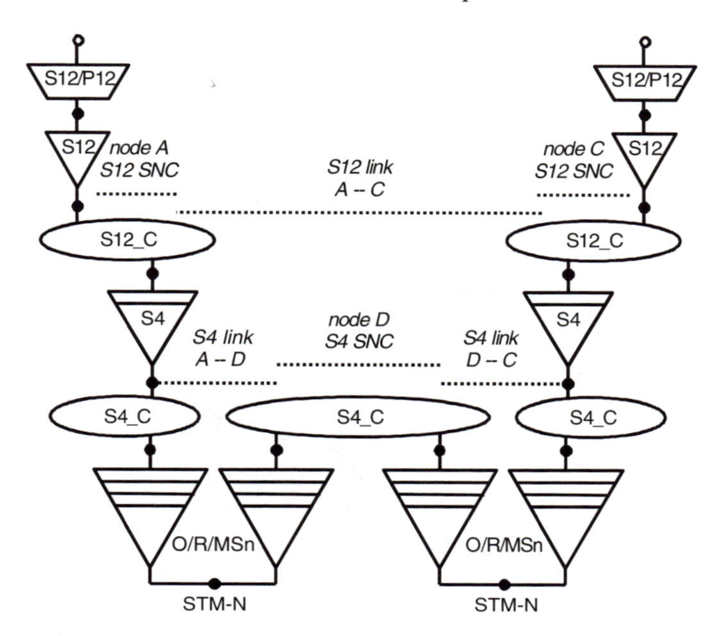

Figure 11.21 Example connection model with server layer.

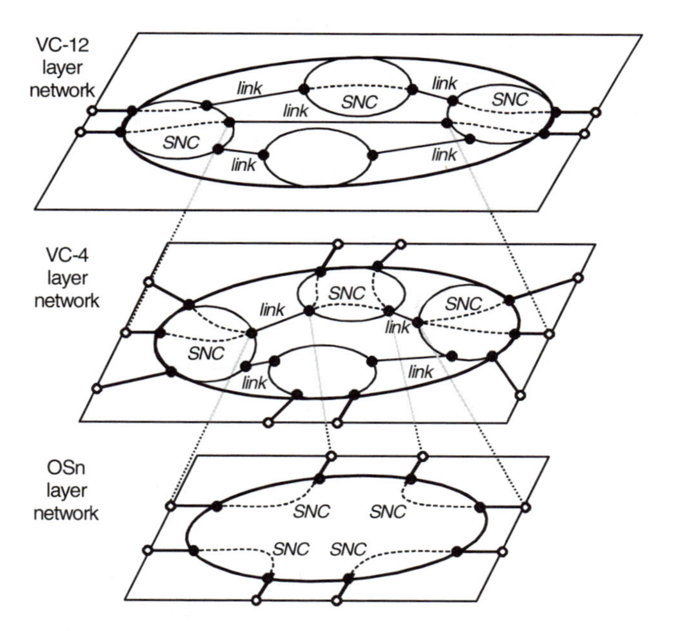

Figure 11.22 Abstract ring model with three layers.

network represent the transport capacity of a VC–4 and could also be represented by 63 individual VC–12 link connections. Similarly, the links in the VC–4 layer network represent the transport capacity of an STM-N and can be replaced by N separate VC–4 link connections. The sub-network connections in the OSn layer network represent the STM-N connectivity in the SDH ring.

Sometimes even this level of detail is not sufficient, especially in those cases where divers routing is important to provide a desired availability. The STM-N signal is transported by a single fiber; this fiber can be bundled with other fibers at a distribution frame. In this way a fiber carrying the working signal can be bundled with the fiber carrying the protection signal in the same cable. It is also possible that in the outside plant a cable containing a fiber carrying the working signals is laid in the same trench or duct as another cable containing a fiber carrying the protection signal. The worst case is when a bridge may have two cable ducts, one on the right side and one on the left; however, this problem can be resolved by considering all cable ducts in a bridge to be a single duct.

Now, in the example functional models provided in this section, two extra layers can be added using similar atomic or compound functions: a cable layer and a duct/trench layer (see Figure 11.23).

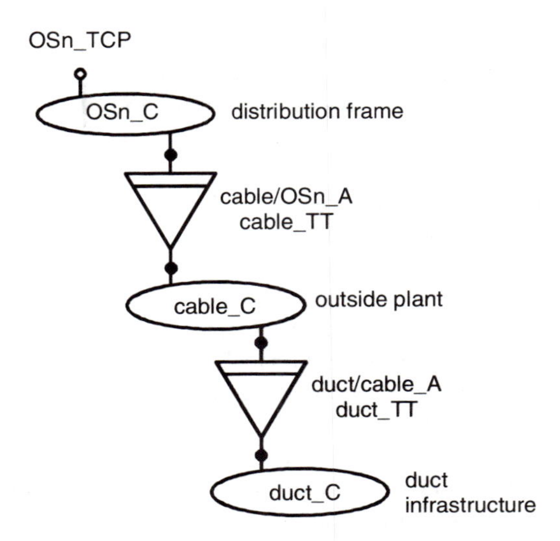

Figure 11.23 Example of cable and duct layers.

11.5 SYNCHRONIZATION NETWORK MODEL

This section provides an example of the network architecture that is used for the distribution of timing information within an SDH network. See ITU-T Rec. G.803 (2000) for a more detailed description. The synchronization functions are also described in ITU-TRec. G.781 (1999).

11.5.1 Synchronization methods

According to ITU-T Rec. G.810 (1996) there are two fundamentally different methodologies that can be used to synchronize the clock of the network elements in an SDH network:

- master-slave synchronization;
- mutual synchronization.

Since master-slave synchronization is the most appropriate methodology for synchronizing SDH networks, only this method is used in the provided example.

In a network supporting master-slave synchronization layers can be distinguished. In the hierarchical structure of a synchronization network the clocks of the network elements in each layer, i.e. each level of the hierarchy, are synchronized by using a reference signal received from a higher level. The highest level in the network is the PRC. The reference signals that are used to synchronize the node clocks are distributed between the levels of the hierarchy. Normally the reference signals are distributed by using the facilities of the transport network. The hierarchical levels are defined in the following recommendations:

- Highest level, primary reference clock: ITU-T Rec. G.811 (1997).
- Slave clock: ITU-T Rec. G.812 (2004).
- SDH network element clock: ITU-T Rec. G.813 (2003).

In an SDH network that distributes the synchronization reference signals only the STM-N signal itself can be used to transport the timing information. Any signal transported in an STM-N structure is subject to pointer adjustments. The unpredictable nature of pointer adjustments severely affects the quality of the downstream slave clock.

The master-slave method is a uni-directional synchronization technique where the slave clock process determines which synchronization trail will be selected as its reference and which alternative synchronization trail will be switched to if the original trail fails.

11.5.2 Synchronization network architecture

The architecture employed in an SDH synchronization network requires that the timing of all network element clocks shall be traceable to a *Primary Reference Clock* (PRC) that is compliant with ITU-T Rec. G.811 (1997) and has a good short-term stability performance in order to comply with the generic slip rate objectives in ITU-T Rec. G.822 (1988). Figure 11.24 shows the distribution of the inter-station synchronization information and the hierarchical relationship between the

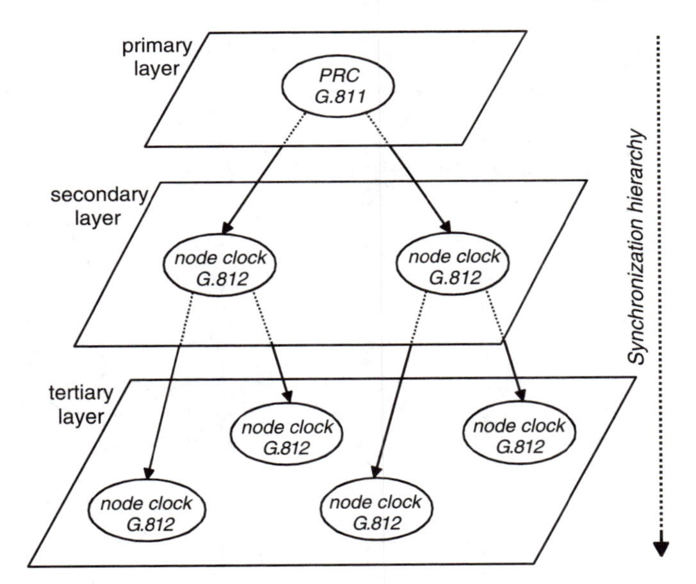

Figure 11.24 Synchronization network hierarchy layering.

nodal clocks. This distribution enables all the stations in the SDH network to be synchronized by utilizing a tree-like topology. During the development and while deploying this architecture, it is important to ascertain—for the correct operation of the synchronization network—clocks of a lower hierarchical level shall only accept timing information from clocks of the same or higher hierarchical level and that timing loops are avoided. To make sure that these requirements are complied with under all circumstances, the distribution network architecture and its clock processes shall be designed so that, even under fault conditions, only valid higher level references are presented to hierarchical clocks at the same or lower levels.

The most important objective of designing the architecture of synchronization networks concerns the modeling of the transport of timing information between the hierarchical layers' synchronization clocks. An example functional model is provided in Figure 11.25. This figure shows three layer networks:

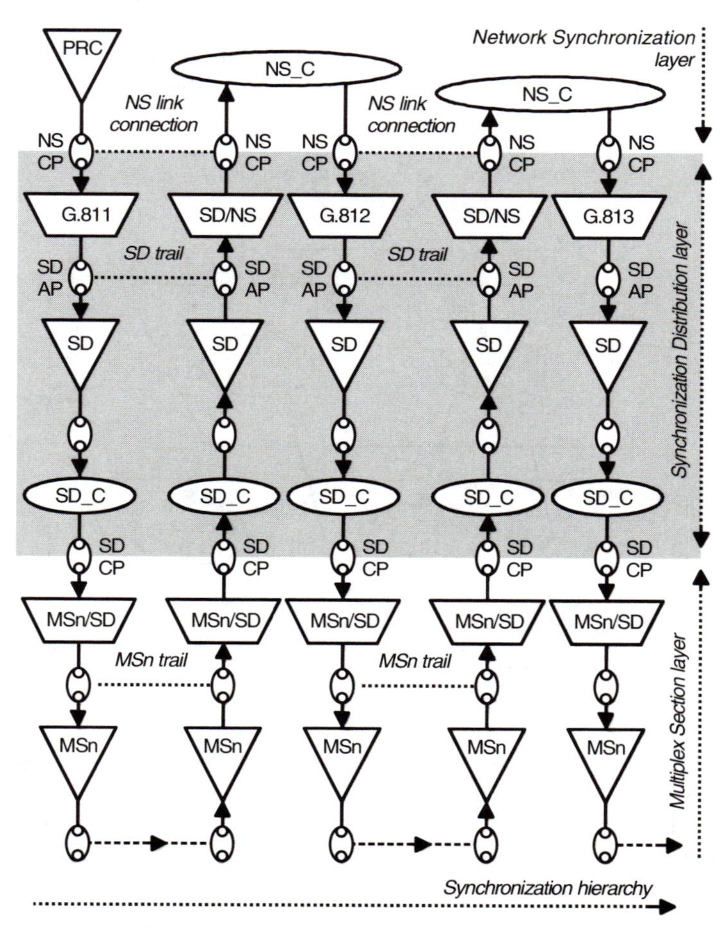

Figure 11.25 Example of synchronization network functional model.

- *Network Synchronization* (NS) layer network. This is the client layer in the functional model. The only responsibility of the NS layer is to provide the point-to-point or point-to-multipoint connections of the PRC to all other node clocks in the network. An estimate of *Universal Time Coordinated* (UTC) is available at every connection point in the NS layer (NS_CP). The quality of this estimate of UTC depends on the provisioning of the NS layer network connection functions and the timing quality of the server trails provided by the server layer network.

  The connection matrices NS-C shown in the NS layer are used to provision the configuration of the synchronization network. Autonomous reconfiguration of the synchronization network, including protection switching similar to the sub-network connection protection scheme described in Chapter 9, Section 9.2.2, is also performed via these matrices.

  The NS link connections between the matrices represent the information flow provided by the SD trails in the SD layer.
- *Synchronization Distribution* (SD) layer network. This is the server layer network of the NS layer network. All network synchronization clocks are located in this layer. The SD layer network provides the trails for the transfer of timing information from one synchronization clock to another in hierarchical order. The transfer of timing information is uni-directional; therefore, all access points of the SD layer network (SD_AP) are uni-directional. The SD layer may be the client layer of any server layer provided that these server layers are transparent for timing information.

  There are three synchronization clocks respectively defined in ITU-T Rec. G.811 (1997), G.812 (2004) and G.813 (2003). The clocks are represented in the functional model by adaptation functions that modify the quality of the timing information according to their specific quality level.

  The connection matrices SD_C in the SD layer are for the provisioning of the SD trails. They are used to select the server layer multiplex sections or paths that will transport the SD trails.

  For the purpose of network performance *Synchronization Status Messaging* (SSM) may be used to convey timing quality information (see ITU_T Rec. G.707 (2003)). SSM messages convey clock source quality level information that may assist clocks to select the most suitable synchronization reference from the set of available references. This information is inserted at the SD trail termination source and extracted at the SD trail termination sink. The SD trail termination sink function also reports the failure of an SD trail towards the NS connection function NS_C to initiate protection switching.

- *Multiplex Section* (MS) layer network. Just for this example the SDH MS layer network is used. However, it may be any layer network, e.g. any multiplex section layer or any path layer, provided that it is transparent for timing information. The impact that pointer adjustments have on the transported timing information means that SDH path layers (e.g. VC–4) and PDH path layers (e.g. P12) that are supported by SDH path layers do not qualify as such.

Note that line system regenerator clocks are contained in the server layers of the SD layer. These regenerator clocks, and in general all clocks in the SD server layers, shall be synchronization timing 'transparent'. The regenerator clocks either transfer timing or squelch the timing information. Conversely, the SD clocks always provide timing information even in case of a failure of one of the preceding SD trails transferring the timing information from the previous clock in the NS link connection.

11.6 OTN NETWORK ELEMENT MODEL

In Figure 11.3 of Section 11.2, a network element is shown that already partially provides OTN functionality. In this section, an example of a functional model of an OTN ADM/MSAP will be described. At the top of Figure 11.26, the tributary inputs are shown as the client layer connection points that convey the client characteristic information *client_CI*. This client signal is adapted for the transport in the OTN network by the function ODUkP/client_A.

Several client signals are supported:

- RSn_CI, the characteristic information of the SDH Regenerator Section as defined in ITU-T Rec. G.783 (2004) clause 10.
- CBRx_CI, a *Constant BitRate* signal with approximate bitrate x, e.g. x = 2G5 or 10G, as defined in ITU-T Rec. G.798 (2004) clauses 12.3.3 and 14.3.1.
- ATM_CI, an ATM signal as defined in ITU-T Rec. I.732 (2000) Annex D.
- PRBS_CI, a *Pseudo Random Bit Sequence* test-signal of length $(2^{31} - 1)$ bits as defined in ITU-T Rec. O.150 (2002) clause 5.8.

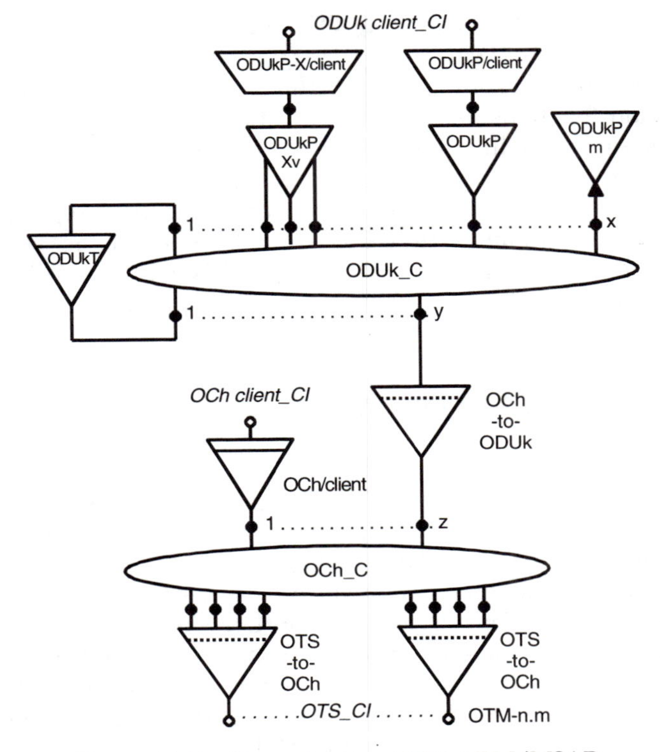

Figure 11.26 Example of an OTN ADM/MSAP.

- NULL_CI, an all-zero bits test-signal (see ITU-T Rec. G.709 (2003) clause 17.4.1).

The ODUkP termination function provides the performance monitoring of the ODUKP trail through the OTN. Note that the OTN also supports transport of concatenated ODUk signals, i.e. ODUk–X_CI, to provide more flexibility in transported payload bandwidth. The actual transport utilizes the same inverse multiplexing methodology that is used in SDH, i.e. *Virtual Concatenation* (VCAT) (see Chapter 10 for more detail).

The ODUk–Xv_TT function provides the VCAT capability. The ODUkPm_TT_Sk function provides performance monitoring capabilities that can be used for the support of non-intrusive protection switching SNC/I (see Chapters 9, Section 9.2.2 for more details). The ODUkPT compound function provides TCM capabilities similar to those of SDH TCM, however, in OTN several levels of TCM can be nested.

The description of the ODUk_C function and the adaptation and termination functions between this function and the OTN OTM–n.m signal is provided in Section 11.3 of this chapter. The connection function OCh_C provides the connectivity of optical signals in this network element. The compound function OCh/client represents the OCh_TT function and the OCh/client_A function. Several client signals are supported; the RSn_CI and CBRx_CI are similar to the ODUk adaptation. The support of a GbE mapping into an OCh is currently under study.

11.7 DATA TRANSPORT MODEL

The most recent addition to the capabilities of SDH and OTN is data packet transport. Contrary to TDM signals, which are connection oriented, data signals generally have a connectionless nature. To enable the modeling of data signal transport, specific functional models and related combination rules have been defined (see ITU-T Rec. G.809 (2003)). For the transport of data signals in SDH or OTN networks they are mapped in containers, a function performed by adaptation functions that are defined for this purpose. The most common mapping mechanism is the *Generic Framing Procedure* (GFP). After the introduction of the MSPP it was also necessary to model the additional data services, e.g. Ethernet and MPLS. Currently, a suite of ITU-T recommendations is being developed covering the full functionality of Ethernet transport network architecture and equipment that describe the atomic functions and the functional equipment models for these additional capabilities, i.e. ITU-T Recommendations G.8010 (2004), G.8011 (2004), G.8011.1 (2004), G.8012 (2004), G.8021 (2004) for Ethernet and G.8110 (2005) and G.mplseq (2005), still under study, for MPLS.

In the next sections, examples are provided for GFP equipment models, Ethernet layer network models and MPLS layer network models.

11.7.1 Equipment models for GFP

GFP mapping can be deployed in transport network elements (e.g. SDH) and in data network elements (e.g. IP, Ethernet Routers). Since

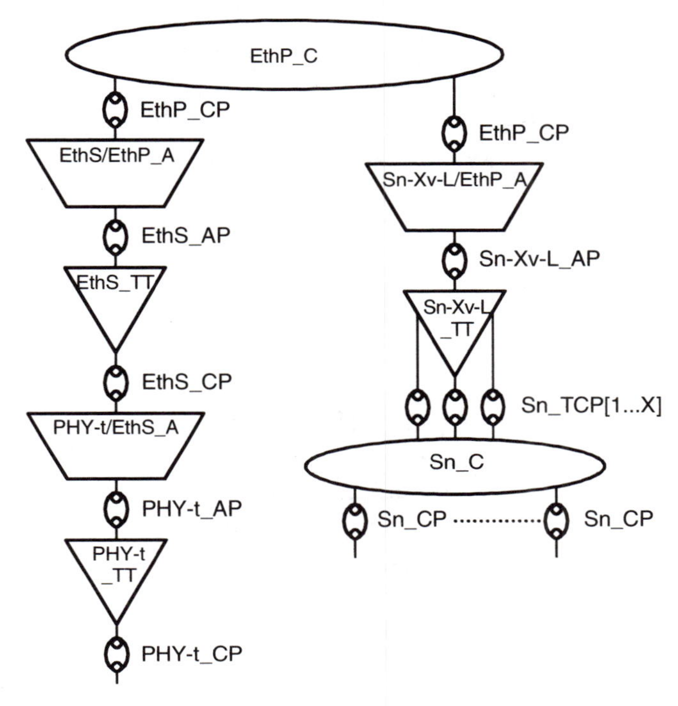

Figure 11.27 Ethernet signal transport using GFP-F.

GFP is a mapping methodology it is found in the application specific functional model of the adaptation function that maps a PDU based client signal into an ordinary signal Sn, a virtual concatenated signal Sn–Xv or a contiguous concatenated signal Sn–Xc. As GFP provides the mapping of a multitude of client signals, a selection has been made to give some examples of the GFP functional model.

Figure 11.27 shows the functional model of an Ethernet tributary port on a transport network element. Figure 11.28 depicts the functional model where GFP processing is performed in an IP Router, or an IP router function embedded in a hybrid SDH/IP network element. Figure 11.29 shows the functional model of a SAN (8B/10B coded signal) tributary port on a transport network element. All figures provide the bi-directional functional model. To avoid confusion, the atomic functions are described for a signal flow from left to right; in the opposite direction, the same atomic functions will perform the reverse operation.

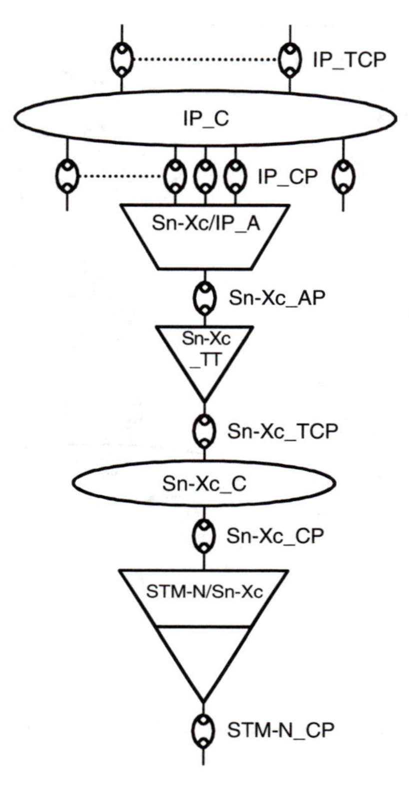

Figure 11.28 IP router interface.

11.7.2 Ethernet tributary port

This example shows a case where only a part of the physical interface bandwidth is carrying traffic and only this traffic is to be transported through the transport network. This is depicted in Figure 11.27. The physical data interface signal is terminated and the data PDUs are extracted. GFP-F mapping into an Sn, Sn–Xc, or Sn–Xv signal is used to transport the data PDUs in the SDH/SONET network.

- The termination function PHY–t_TT terminates the physical Ethernet signal with its characteristic information present at the (physical) connection point PHY–t_CP.
- The adaptation function PHY–t/EthS_A provides the Ethernet PHY decoding and data/clock recovery.

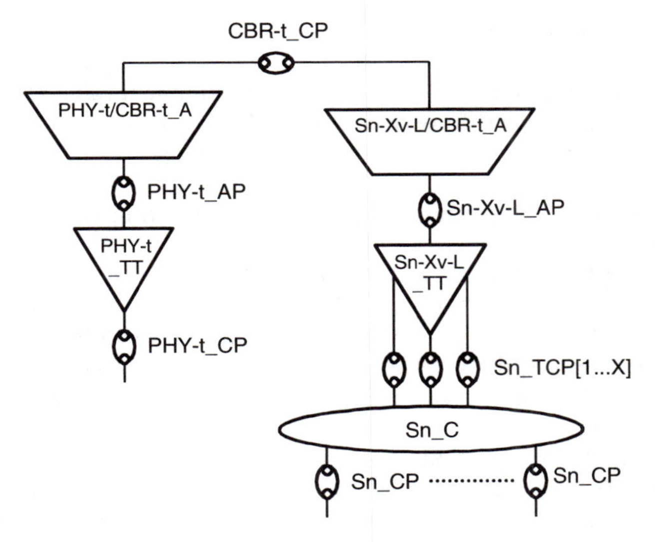

Figure 11.29 SAN signal transport using GFP-T.

- The termination function EthS_TT processes the Ethernet control characters.
- The adaptation function EthS/EthP_A extracts the Ethernet MAC frames or PDUs, i.e. the Ethernet characteristic information.
- The connection function EthP_C is used to transfer the signal from tributary port to the network port for further transport. It has the same functionality as an Ethernet bridge or switch function in a router.
- The adaptation function Sn–Xv–L/EthP_A encapsulates the Ethernet MAC frames into GFP-F frames and maps the GFP stream into a virtual concatenated signal with LCAS enabled Sn–Xv–L to provide flexibility and protection. Other possibilities would have been to map the GFP stream into a ordinary signal Sn, a contiguous concatenated signal Sn–Xc, or a virtual concatenated signal with LCAS disabled Sn–Xv.
- The termination function Sn–Xv–L_TT provides the termination of the X Sn network connections, i.e. the equivalent of an Sn–Xv–L network connection.
- The connection function Sn_C provides the connectivity for the X virtual concatenation member signals Sn_CI.

11.7.3 IP router port

In this example, only the part of the functional model where GFP mapping is performed in between the IP Router fabric and the STM-N interface port functions is shown in Figure 11.28.

- The IP connection function is the representation of an IP router with many ports.
- The adaptation function Sn–Xv–L/IP_A provides the encapsulation of IP frames into PPP frames, the encapsulation of PPP frames into PPP/HDLC frames, the PPP/HDLC frames into GFP-F frames and the mapping of the GFP stream into an Sn–Xc signal.
- The termination function Sn–Xc_TT provides the Sn–Xc signal termination.
- The connection function Sn–Xc_C is either a fixed binding between the Sn–Xc signal TCP and CP, or is provisionable to connect several GFP streams to one or more STM-N interfaces present on the IP router.
- The compound function STM-N/Sn–Xc represents the atomic functions required to adapt the Sn–Xc signal for transport over the physical network connection between the connection points STM–N_CP.

11.7.4 SAN tributary port

This example shows the functional model for those cases where the physical data signal is an 8B/10B coded signal. This signal can be transported through the transport network as a transparent stream using GFP-T mapping. Figure 11.29 illustrates the functional model from the physical interface to the first connection function.

- The termination function PHY–t_TT terminates the physical SAN signal presented at the (physical) connection point PHY–t_CP, e.g. Fibre Channel, ESCON, FICON or GbEthernet signals.
- The adaptation function PHY–t/CBR–t_A provides the 8B/10B decoding and data/clock recovery. The output is a constant bitrate signal CBR–t _CI.
- The adaptation function Sn–Xv–L/CBR–t_A encapsulates the PHY layer characters (data and control) into fixed-length GFP-T frames

and maps the GFP stream into a virtual concatenated signal with LCAS enabled Sn–Xv–L. The Sn–Xv–L provides protection by provisioning more member bandwidth than required by the CBR–t signal. Other possibilities would have been to map the GFP stream into a ordinary signal Sn, a contiguous concatenated signal Sn–Xc, or a virtual concatenated signal with LCAS disabled Sn–Xv; in these cases protection can be provided by using SNCP (see Chapter 9, Section 9.2.2).
- The termination function Sn–Xv–L_TT provides the termination of the X member signals Sn_CI.

The connection function Sn_C provides the connectivity for the X member signals Sn_CI of the virtual concatenation group.

11.8 ETHERNET LAYER NETWORK MODEL

For Ethernet, two layer networks are defined in the Ethernet transport network architecture:

- ETY: the Ethernet PHY (physical) layer network. This is a section layer network and can be the server layer network of:
- ETH: the Ethernet *Media Access Control* (MAC) layer network. This is a path layer network. The characteristic information of the ETH layer network can be transported through ETH links supported by trails in the server layer networks (e.g., ETYn, SDH Sm/Sn, OTN ODUkP, MPLS, ATM).

11.8.1 ETY layer network terminology

In the ETYn layer network, adapted ETH characteristic information is transported through an ETYn trail between ETYn access points. The ETYn characteristic information is the physical section signal that will be transported over the physical medium (e.g. fiber, copper). Specific ETYn signal types are defined in IEEE 802.3 (2002). Examples of those signal types, grouped by rate, are shown in Table 11.1.

For the ETYn layer network, the following topological components, transport entities, reference points and transport processing functions are defined as illustrated in Figure 11.30.

Table 11.1 ETYn signal types.

n	ETYn
1	10BASE set of signals
2	100BASE set of signals
3	1000BASE set of signals
4	10GBASE set of signals

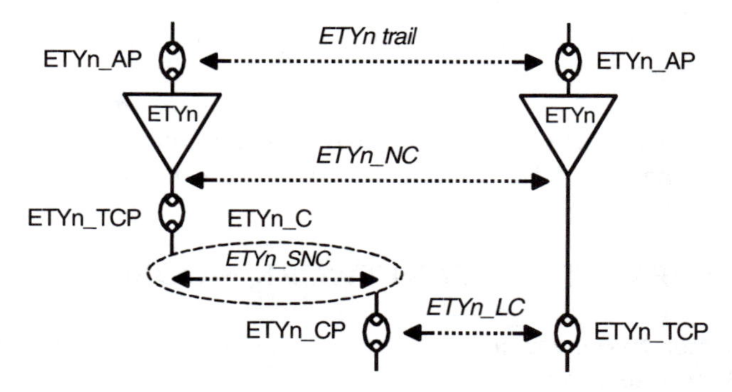

Figure 11.30 Ethernet ETYn layer functional model.

- the ETYn trail between the ETN access points ETYn_AP;
- the ETYn trail termination ETYn_TT that can be decomposed into the uni-directional source function ETYn_TT_So and sink function ETYn_TT_Sk;
- the ETYn link connection ETYn_LC indicating the transport capability between two ETY termination connection points ETYn_TCP or between an ETY connection point ETYn_CP and an ETYn_TCP;
- the ETYn network connection ETY-NC; either a single link connection between two ETYn_TCP or a series of ETYn_LC and sub-network connections ETYn_SNC;
- the ETYn sub-network connection function ETYn_C representing the connectivity for the ETYn characteristic information transported over an ETYn_SNC provided in a server layer network.

11.8.2 ETH layer network terminology

In the ETH layer network, adapted information is transported through a connectionless ETH trail between ETH access points. The ETH

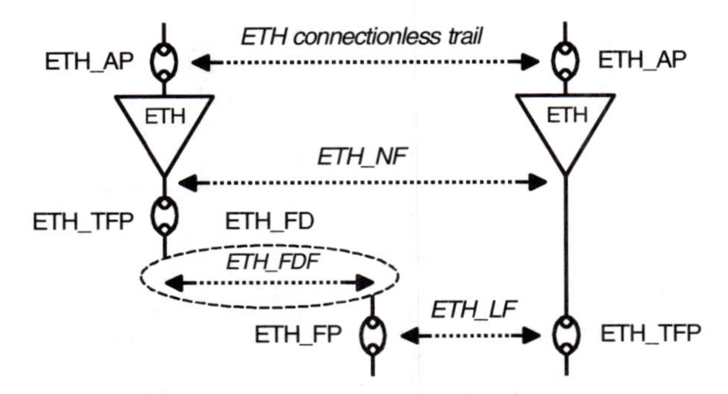

Figure 11.31 Ethernet ETY layer functional model.

adapted information is a flow of MAC Service Data Units, as defined in IEEE 802.3 (2002).

For the ETH layer network, the following topological components, transport entities, reference points and transport processing functions are defined as illustrated in Figure 11.31.

- the ETH connectionless trail between the ETH access points ETH_AP;
- the ETH flow termination function ETH_FT that can be decomposed into the uni-directional source function ETH_FT_So and sink function ETH_FT_Sk;
- the ETH link flow ETH_LF indicating the transport capability between two ETH termination flow points ETH_TFP or between an ETH flow point ETH_FP and an ETH_TFP;
- the ETH network flow (NF); a series of ETH_LF and flow domain flows ETN_FDF;
- the ETH flow domain (FD) representing the connectivity for the ETH characteristic information transported over an ETH_FDF provided in a server layer network.

11.8.3 Ethernet layer network example

Figure 11.32 contains a summary of the atomic functions that may be shown in a functional model of the Ethernet transport layer part of a network model or in an equipment model. It is not necessary that all atomics are present at the same time, especially the ETH to client adaptation functions and the server to ETH termination and

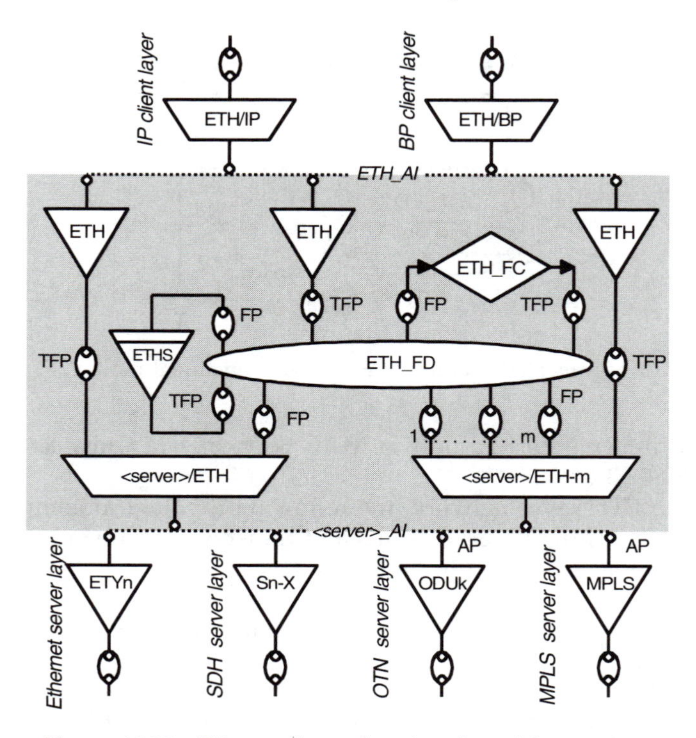

Figure 11.32 Ethernet layer functional model overview.

adaptation functions. A particular configuration can contain a subset of these atomic functions. The Ethernet layer is located between the client layer network, e.g. an *Internet Protocol* (IP) or *Bridging Protocol* (BP) layer and the server layer, e.g. an ETYn, SDH, OTN or MPLS transport layer.

Apart from the functions described in Section 11.8.2, the following functions are shown in the ETH layer:

- the ETH flow conditioning function ETH_FC; the objective is to determine the conformance of the incoming Ethernet frames. If a received Ethernet frame does not conform it is discarded.
- the ETH sublayer (connectionless) trail monitoring compound function (ETHS) represents the possibility of inserting a (nested) sublayer for additional OAM functionality. ETH OAM requirements are defined in ITU-T Rec. Y.1730 (2004).
- the server layer to ETH adaptation function supporting IEEE 802.1D Queuing (2004), <server>/ETH_A.

- the server layer to ETH adaptation function supporting IEEE 802.1Q Queuing (2003), <server>/ETH-m_A.

On top of the ETH layer are the client layers with the ETH/IP_A and ETH/BP_A functions. Below the ETH layer are the server layers. Figure 11.32 shows several possible trail termination functions. The Sn-X_TT function represents the SDH lower order Sm_TT, Sm–Xv_TT and higher order Sn_TT, Sn–Xv_TT and Sn–Xc_TT functions. The ODUk_TT function represents the OTN ODUkP_TT and ODUkP_Xv_TT functions.

It is not required to connect the client layers to the server layers via the ETH_FD function; they can be bound directly without a connection function as indicated at the right side of the figure.

11.9 MPLS LAYER NETWORK MODEL

The *Multi-Protocol Label Switching* (MPLS) layer network is currently under study by the experts of ITU-T in Study Group 15, working party 3. The MPLS architecture is described in the recently published ITU-T Rec. G.8110 (2005) and work is in progress on the equipment specification available as draft ITU-T Rec. G.mplseq (2005). The following sections are provisional and should be used only as an indication of possible definitions.

11.9.1 MPLS layer network terminology

In the MPLS layer network, MPLS adapted information is transported through an MPLS trail between MPLS access points. The MPLS characteristic information is the information that will be transported by the server layer network. For the MPLS layer network the following topological components, transport entities, reference points and transport processing functions are defined as illustrated in Figure 11.33.

- the MPLS trail between the MPLS access points MPLS_AP;
- the MPLS trail termination MPLS_TT that can be decomposed into the uni-directional source function MPLS_TT_So and sink function MPLS_Sk;
- the MPLS link connection MPLS_LC indicating the transport capability between two MPLS termination connection points

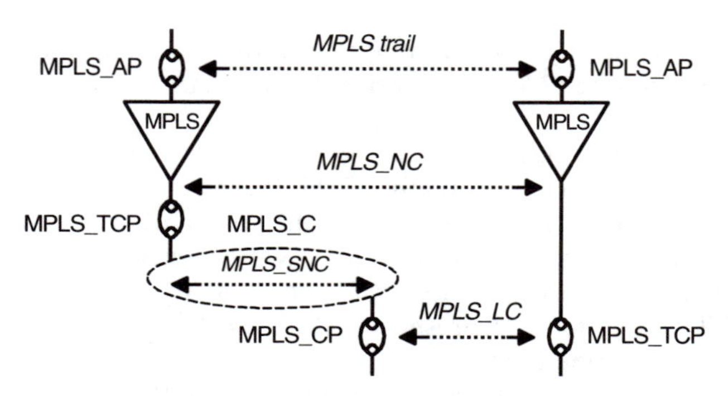

Figure 11.33 MPLS layer terminology.

MPLS_TCP or between an MPLS connection point MPLS_CP and an MPLS_TCP;

- the MPLS network connection MPLS-NC; either a single link connection between two MPLS_TCP or a series of MPLS_LC and sub-network connections MPLS_SNC;
- the MPLS sub-network connection function MPLS_C representing the connectivity for the MPLS characteristic information transported over an MPLS_SNC provided in a server layer network.

11.9.2 MPLS layer network example

Figure 11.34 contains a summary of the atomic functions that are under construction. After the definitions are settled they may be shown in a functional model of the MPLS transport layer. It is not necessary that all atomics are present at the same time, especially the MPLS to client adaptation functions and the server to MPLS termination and adaptation functions. A particular configuration may contain a subset of these atomic functions. The MPLS layer is located between the client layer network, e.g. *Ethernet* (ETH), *Internet Protocol* (IP), *Frame Relay* (FR) and/or *ATM* VC layer and the server layer, e.g. ETH, SDH and/or OTN transport layers.

Apart from the generic MPLS atomic functions shown in Figure 11.34, the following optional functions are shown in the MPLS layer:

- the Ethernet to MPLS interworking function ETH⟨⟩MPLS;
- the ATM to MPLS interworking function VC⟨⟩MPLS;
- the MPLS non-intrusive monitoring function MPLSm_TT;

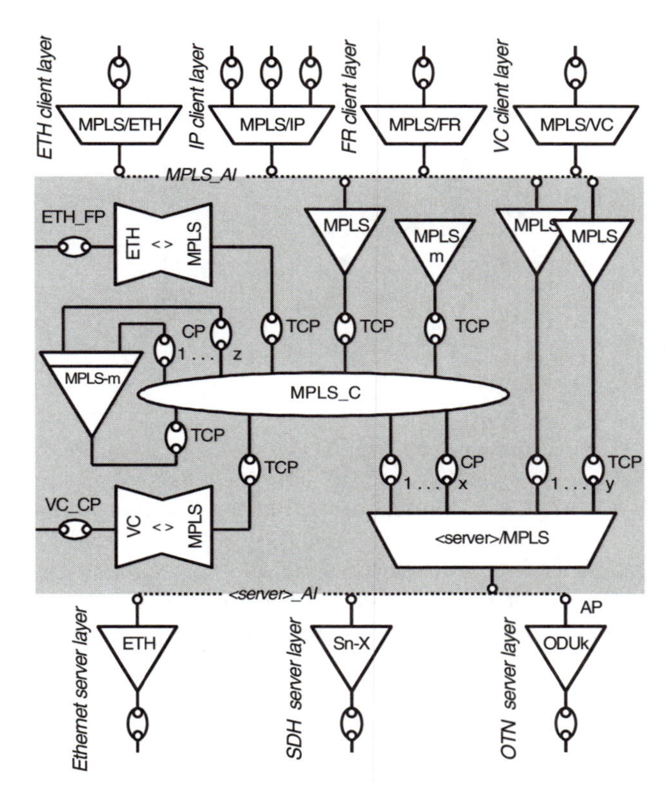

Figure 11.34 MPLS layer functional model overview.

- the compound function MPLS-m that can provides embedded hierarchy in the MPLS layer by utilizing the label stacking mechanism. In the model it represents the MPLS tunnel sublayers and contains an MPLS_TT and MPLS/MPLS-m_A function.

It is not required to connect the client layers to the server layers via the MPLS_C function; they can be bound directly without a connection function as is indicated at the right side of the figure.

Below the MPLS layer are the non-MPLS server layers. Figure 11.34 shows several possible trail termination functions. The ETH_TT function, the Sn–X_TT function that represents the possible SDH lower order Sm_TT, Sm–Xv_TT and higher order Sn_TT, Sn–Xv_TT and Sn–Xc_TT functions and the ODUk_TT function that represents the OTN ODUkP_TT and ODUkP_Xv_TT functions.

Glossary

Terms and definitions

The following conventions are used in this section of the book:

- The terms used in this book are specific to the G and Y series of ITU-T Recommendations, and should not be confused with the same terms used in other Recommendations (e.g. the I series).
- Where a description or definition of a particular term contains another term that is itself defined in this section, that term is given in *italics*.
- The identifying qualifiers of the terms used in the functional models are shown in square brackets (e.g. [_TT_So]).
- The terms can be further qualified by reference to a specific layer network by adding the appropriate layer network qualifier (e.g. SDH higher-order Sn *path termination*, PDH 2 048 kbit/s *path termination*, Ethernet MAC *path termination*).
- All architectural components are bi-directional unless qualified by the term sink or source or uni-directional.

SDH/SONET Explained in Functional Models Huub van Helvoort

Term	Description/definition
access group	In a connection oriented layer network: A group of co-located *trail termination functions* together with their associated *access points* that are connected to the same *sub-network* or *link*. Generally bi-directional. In a connectionless layer network: A group of co-located *flow termination functions* together with their associated *access points* that are attached to the same *flow domain* or *flow point pool link*. Generally uni-directional.
access point	**[_AP]** A *reference point* that: In a connection oriented layer network, consists of an associated contra directional pair of co-located *uni-directional access points*. It represents the *binding* between the *trail termination function* and the *adaptation function*. In a connectionless layer network, consists of a *uni-directional access* point. It represents the binding between the *flow termination function* and the *adaptation function.* An access point is characterized by the *adapted information* that passes across it.
adaptation (function)	**[_A]** A *transport processing function* that adapts a *server layer* to the needs of a *client layer*. It consists of an associated and co-located *adaptation source* and *adaptation sink* pair. It defines the *client/server relationship* between the *connection point* and the *access point*. It is an atomic function that passes a collection of information between layer networks by changing the way in which the collection of information is presented.
adaptation sink	**[_A_Sk]** A *transport processing function* that accepts the *adapted information* presented at its input by the *server layer* network *trail termination sink* and after processing presents the *client layer* network *characteristic information* at its output.
adaptation source	**[_A_So]** A *transport processing function* that accepts the *client layer* network *characteristic information* received at its input and prepares it for the transport over a *trail* in the *server layer* network. It outputs *adapted information* to the *trail termination source* function.

Term	Description/definition
adapted information	**[_AI]** A signal that is transferred on *trails*. The format is technology specific and is defined in the appropriate recommendations.
administrative domain	A domain that represents the set of resources that belong to a single owner such as a network operator, a service provider or an end-user. Administrative domains of different owners do not overlap amongst themselves.
administrative unit	(AU) The *information structure* that provides the *adaptation* between the higher order *path layer* and the multiplex *section layer*. It consists of an information payload (the higher order VC) and an AU pointer that indicates the offset of the payload frame start relative to the multiplex section frame start.
architecture	A description of the structure and relationship among *architectural components* (hardware and software) of a system. The system architecture may include the system's interface with its management environment.
architectural component	Any item required to provide a generic description of the functionality of a transport network independent of the implementation technology.
atomic function	A function that, if divided into simpler functions, would cease to be uniquely defined for transport networks. Hence it is indivisible from a network point of view. The following atomic functions are defined in each network layer: - *connection function* - *adaptation function* - *trail termination function*
bi-directional access point	**[_AP]** The formal term for an *access point*.
bi-directional connection	**[_C]** The formal term for a *connection*.
bi-directional connection point	**[_CP]** The formal term for a *connection point*.
bi-directional port	The formal term for a *port*.

Term	Description/definition
bi-directional termination connection point	**[_TCP]** The formal term for a *termination connection point*.
bi-directional trail	The formal term for a *trail*.
binding	A direct relationship between a *transport processing function* or a *transport entity* and another *transport processing function* or *transport entity* that represents a static connection that cannot be directly modified by a management action.
characteristic information	**[_CI]** A signal of a specific rate and format that is transferred within and between *sub-networks* and is presented to an *adaptation* function for transport by the *server* layer network. The format is technology specific and is defined in the appropriate recommendations.
circuit	This may be regarded as a *trail* in the *circuit layer network*.
circuit layer network	A *layer network* that is concerned with the transfer of information between circuit layer *access points* in direct support of telecommunication services.
client layer network	A *transport network layer* using the transport capability of the *transport network layers* below it.
client/server layer	Any two adjacent *network layers* that are associated in a *client/server relationship*. Each *transport network layer provides* transport capability to the *layer network* above it and uses the transport capability of the *layer networks* below. The layer providing transport is termed a server; the layer using transport is termed client.
client/server relationship	The association between adjacent *layer networks* that is performed by an *adaptation function* to allow the *link connection* in the *client layer network* to be supported by a *trail* in the *server layer network*.
compound function	A function that represents a collection of *atomic functions*. Examples of compound functions: - A combination of several atomic adaptation functions within a certain layer (each serving one client layer) is a compound adaptation function. - A combination of a (compound) adaptation function and the layer's termination function. See also *major compound functions*.

Term	Description/definition
connection	A *transport entity* that consists of an associated pair of co-located *uni-directional connections* capable of simultaneously transferring information in opposite directions between their respective inputs and outputs.
connection function	**[_C]** An *atomic function* within a layer that, if connectivity exists, relays a collection of items of information between groups of *atomic functions*. It does not modify the members of this collection of items of information although it may terminate any switching protocol information and act upon it. Any connectivity restrictions between inputs and outputs are stated.
connection point	**[_CP]** A *reference point* that consists of an associated contra directional pair of co-located *uni-directional connection points*. It represents the *binding* of a *connection function* and an *adaptation function*. A connection point is characterized by the *characteristic information* that passes across it.
connection supervision	The process of monitoring the integrity of a *connection* or *tandem connection* that is part of a *trail*.
connectionless trail	A *transport entity* responsible for the transfer of information from the input of a *flow termination source* to the output of a *flow termination sink*. The integrity of the information transfer may be monitored.
consolidation	The allocation of *server layer trails* to *client layer connections*, which ensures that each *server layer trail* is full before the next is allocated. Consolidation minimizes the number of partially filled *server layer trails*. It therefore maximizes the fill factor. Thus a number of partially filled VC–4 paths may be consolidated into a single fully filled VC–4.
dedicated protection	A protection architecture that provides capacity dedicated to the protection of traffic-carrying capacity (1 + 1).
dual ended operation	A protection operation method that takes switching action at both ends of the protected entity (e.g. *trail*, *sub-network connection*), even in the case of a uni-directional failure.
equipment functional specification	A collection of *atomic, compound, and/or major compound functions* and overall performance objectives that describe the functionality of equipment.

Term	Description/definition
fault	A fault is the inability of an item to perform a required function, excluding that inability due to preventive maintenance, lack of external resources or planned actions. *Note*: A fault is often the result of a failure of the item itself, but may exist without prior failure.
flow	An aggregation of one or more *traffic units* with an element of common routing.
flow domain	**[_FD]** Is an *atomic function* within a layer that, if connectivity exists, relays a collection of items of information between groups of *atomic functions*. It does not modify the members of this collection of items of information although it may terminate any switching protocol information and act upon it. Any connectivity restrictions between inputs and outputs are stated.
flow domain flow	(FDF) A *transport entity* that transfers information across a *flow domain*, it is formed by the association of *ports* on the boundary of the *flow domain*.
flow point	**[_FP]** A uni-directional *reference point* that represents the *binding* of a *flow domain* and an *adaptation function*. A flow point is characterized by the *characteristic information* that passes across it.
flow point pool	**[_FPP]** A group of co-located *flow points* that have a common routing.
flow point pool link	A *topological component* that describes a fixed relationship between a *flow domain* or *access group* and another *flow domain* or *access group*.
flow termination	The flow termination defines the association between the *access point* and the *termination flow point;* these points therefore delimit the trail termination. The *flow termination* is generally uni-directional. A trail termination generates the *characteristic information* of a *layer network* and ensures the integrity of that *characteristic information*.
flow termination sink	**[_FT_Sk]** A *transport processing function* that accepts the *characteristic information* of the *layer network* at its input, recovers the information related to *connectionless trail* monitoring and presents the remaining information at its output to the *adaptation sink* function.

Term	Description/definition
flow termination source	**[_FT_So]** A *transport processing function* that accepts adapted *characteristic information* from a client layer network at its input, adds information to allow the *connectionless trail* to be monitored and presents the characteristic information of the layer network at its output(s).
higher order path	An SDH specific *path layer network*. The higher order (HO) path layers provide a server network for the lower order (LO) path layers. The comparative terms 'lower' and 'higher' refer only to the two participants in such a client/server relationship. VC–11, VC–12 and VC–2 paths may be described as 'lower order' in relation to VC–3 and VC–4, while the VC–3 path may be described as 'lower order' in relation to a VC–4.
interworking function	**[_I]** A *transport processing function* that adapts the *characteristic information* of one *server layer* to the *characteristic information* of another *server layer* provided that the function is transparent for the payload information. It defines the *interworking relationship* between two *connection points*. It is an atomic function that passes a collection of information between server layers by changing the way in which the collection of information is presented.
layer	A concept used to allow the transport network functionality to be described hierarchically as successive levels, each layer being solely concerned with the generation and transfer of its characteristic information. The layering concept of the transport network allows: - each layer network to be described using similar functions; - the independent design and operation of each layer network; - each layer network to have its own operations, diagnostic and automatic failure recovery capability; - the possibility of adding or modifying a layer network without affecting other layer networks from the architectural point of view; - simple modeling of networks that contain multiple transport technologies.

Term	Description/definition
layer network	A *topological component* that includes both *transport entities* and *transport processing functions* that describe the generation, transfer and termination of particular *characteristic information*. It is defined by the complete set of *access groups* of the same type that may be associated for the purpose of transferring information. The associations of the *trail terminations* that form a trail in a layer network may be made and broken by a layer network management process thus changing its connectivity. A separate, logically distinct layer network exists for each termination type. The topology of a layer network is described by access groups, sub-networks, and the links between them.
link	A *topological component* that describes a fixed relationship between a *sub-network* or *access group* and another *sub-network* or *access group*.
link connection	(LC) A *transport entity* that transfers information between *ports* across a *link*. It is provided by a client/server association. It is formed by a near-end *adaptation function*, a server *trail* and a far-end *adaptation function* between *connection points*. It can be configured as part of the trail management process in the associated *server layer*.
link flow	(LF) A *transport entity* that transfers information between *ports* across a *flow point pool link*.
major compound function	A *compound function* that represents a collection of *atomic functions* within more than one *layer*. An example of a major compound function is: - The atomic functions in the Optical Section, Regeneration Section and the Multiplex Section layers may be combined to form a major compound function. (Major) compound functions facilitate simplified descriptions of equipment and network models.
management domain	A management domain defines a collection of managed objects that are grouped to meet organizational requirements according to geography, technology, policy or other structure, and for a number of functional areas such as configuration, security, (FCAPS), for the purpose of providing control in a consistent manner. Management domains can be disjoint, contained or overlapping. As such the resources within an administrative domain can be distributed into several possible overlapping management domains. The same resource can therefore belong to several management domains simultaneously, but a management domain shall not cross the border of an administrative domain.

Term	Description/definition
management information	**[_MI]** Information flow between the management layer and an atomic function, containing information required to provision the atomic function and information retrieved from the atomic function regarding the performance of the processed information, e.g. LCASenable and SF detected.
management point	**[_MP]** A *reference point* where the output of an atomic function is bound to the input of the element management function and/or where the output of an element management function is bound to the input of a atomic function. A management point is characterized by the *management information* that passes across it.
matrix	A topological component used to effect information routing and management. It describes the potential for matrix connections across the matrix. A matrix is contained within one physical node. As such it represents the limit to the recursive partitioning of a *sub-network* or *flow domain*.
matrix connection	A *transport entity* that transfers information across a *matrix*. It is formed by the association of *ports* on the boundary of the *matrix*. It is a *sub-network connection* that may be configured as part of the trail management process or it may be fixed.
matrix flow	A *transport entity* that transfers information across a *matrix*, it is formed by the association of *ports* on the boundary of the *matrix*.
multiplex section layer network	A *layer network* that may be media dependent and which is concerned with the transfer of information between multiplex section layer *access points*.
network	All of the entities (such as equipment, plant, facilities) that together provide communication services.
network connection	(NC) A *transport entity* formed by a series of contiguous *link connections* and/or *sub-network connections* between *termination connection points*.
network flow	(NF) A *transport entity* formed by a series of contiguous *flow domain flows* and/or *flow point pool links* between *termination flow points*.
network node interface	(NNI) The interface at a network node that is used to interconnect with another network node.

Term	Description/definition
pairing	A relationship between source and sink *transport processing functions* or two contra directional uni-directional *transport entities* or between *uni-directional reference points* that have been associated for the purposes of bi-directional transport.
path	A *trail* in the *path layer network.*
path layer network	A *layer network* that is independent of the transmission media and that is concerned with the transfer of information between path layer network *access points.*
path termination	A *trail termination* in a *path layer network.*
path termination sink	A *trail termination sink* in a *path layer network.*
path termination source	A *trail termination source* in a *path layer network.*
physical media layer network	A *layer network* that is concerned with the actual optical fiber, metallic pair, or radio frequency that supports the *section layer network.*
point-to-multipoint connection	A *connection* capable of transferring information from a single input to multiple outputs.
port	In a connection oriented layer network: An associated pair of co-located *uni-directional ports* capable of transferring information in opposite directions. In a connectionless layer network: A *uni-directional port* capable of transferring information in one direction. For management purposes, a port represents either a connection termination point or a trail termination point.
protection	See: *dedicated protection, dual ended operation, shared protection, single ended operation.*
reference point	An architectural component, which is formed by the binding between inputs and outputs of *transport processing functions* and/or *transport entities.* It is characterized by the information that passes across it. It is the delimiter of an atomic function.

Term	Description/definition
regeneration section layer network	A *layer network* that may be media dependent and which is concerned with the transfer of information between regenerator section layer *access points*.
remote defect indicator	(RDI) A signal that returns the defect status of the characteristic information received by the trail termination sink function back to the origination trail termination source function. Examples of RDI: Far-End-Receive-Failure (FERF) in SDH, the A-bit in a structured 2048 kbit/s signal.
remote error indicator	(REI) A signal that returns either the exact or the truncated number of error detection code violations (EDCV) within the characteristic information detected by the trail termination sink function back to the origination trail termination source function. Examples of REI: Far-End-Block-Error (FEBE) in SDH, the E-bit in a structured 2048 kbit/s signal.
remote information	**[_RI]** Information flow from the sink side to the source side of the same trail termination function in uni-directional representation, containing information to be transferred to the far end, e.g. RDI and REI.
remote point	**[_RP]** A *reference point* where the output of a sink side atomic function is bound to the input of the associated source side atomic function. A remote point is characterized by the *remote information* that passes across it.
routing	The process whereby a number of *connection functions* within the same layer are configured to provide a *trail* between *trail termination points*.
section	A *trail* in the *section layer network*.
section layer network	A layer network that is concerned with the transfer of information between section layer access points. In SDH the section layer network is divided into the multiplex section network and the regenerator section layer network.
section termination	A *trail termination* in the *section layer network*.
section termination sink	A *trail termination sink* in the *section layer network*.
section termination source	A *trail termination source* in the *section layer network*.

Term	Description/definition
server layer network	The transport network layer providing transport capability to the transport network layers above it.
shared protection	A protection architecture using m protection entities shared amongst n working entities (m:n). The protection entities may also be used to carry extra traffic when not in use for protection.
single ended operation	A protection operation methodology that takes a switching action only at the affected end of the protected entity (e.g. *trail*, *sub-network connection*), in the case of a uni-directional failure.
sublayer	A set of additional transport processing functions and reference points encapsulated within a layer network. It is created by decomposition of transport processing functions or reference points.
sub-network	A topological component used to provide routing of specific *characteristic information* and management in a network. It describes the potential for sub-network connections across the sub-network. It can be partitioned into interconnected sub-networks and links between sub-networks. Each sub-network in turn can be partitioned into smaller sub-networks and links, etc. A sub-network may be contained within one physical node.
sub-network connection	(SNC) A transport entity that transfers *characteristic information* across a *sub-network*, it is formed by the association of *ports* on the boundary of the *sub-network*. It can be configured as part of the *trail management process*.
sub-network connection protection	(SNCP) A protection type that is modeled by a sublayer that is generated by expanding the *sub-network connection point*.
tandem connection	(TC) An arbitrary series of contiguous *link connections* and/or *sub-network connections*.
tandem connection bundle	A parallel set of *tandem connections* with co-located end points.
termination connection point	**[_TCP]** A *reference point* that consists of an associated contra directional pair of co-located *uni-directional termination connection points*. It represents the *binding* of a *trail termination* to a *bi-directional connection*.

Term	Description/definition
termination flow point	**[_TFP]** A uni-directional *reference point* that represents the *binding* of a *flow termination* and a *flow domain*.
timing information	**[_TI]** Information flow between the synchronization layer and an adaptation source or a connection function, containing information required to enable synchronous transport, e.g. Clock and Frame Start.
timing point	**[_TP]** A reference point where an output of the synchronization distribution layer is bound to the input of an adaptation source or connection function, or where the output of an adaptation source is bound to the input of a synchronization distribution layer. A timing point is characterized by the *timing information* that passes across it.
topological component	An architectural component, used to describe the transport network in terms of the topological relationships between sets of reference points within the same layer network. A topological description in terms of these components describes the routing possibilities of the network and hence its ability to support transport entities.
traffic conditioning function	A *transport processing function* that accepts the *characteristic information* of the *layer network* at its input, classifies the *traffic units* according to configured rules, meters each *traffic unit* within its class to determine its eligibility, polices non-conformant *traffic units* and presents the remaining *traffic units* at its output as *characteristic information* of the *layer network*.
traffic unit	An instance of *characteristic information* and a unit of usage.
trail	A *transport entity* that consists of an associated pair of *uni-directional trails* capable of simultaneously transferring *characteristic information* in opposite directions between their respective inputs and outputs. A trail is formed by combining *trail termination functions* and *network connections*. The integrity of the transferred characteristic information is monitored.

Term	Description/definition
trail management process	Configuration of network resources during network operation for the purposes of allocation, re-allocation and routing of *trails* to provide *transport* to client networks.
trail protection	A protection type that is modeled by a sublayer that is generated by expanding the *trail termination*.
trail termination	The trail termination defines the association between the *access point* and the *termination connection point;* these points therefore delimit the trail termination. It consists of a co-located *trail termination source* and *trail termination sink*. A trail termination generates the *characteristic information* of a *layer network* and ensures the integrity of that *characteristic information*.
trail termination function	**[_TT]** An atomic function within a layer that generates, adds, and monitors information concerning the integrity and supervision of adapted information. It is a *transport processing function* that accepts the *characteristic information* of the *layer network* at its input, removes the information related to monitoring the *trail* and presents the remaining information at its output; or it accepts *adapted information* from a *client layer network* at its input, adds information to allow the *trail* to be monitored and presents the characteristic information of the layer network at its output.
trail termination sink	**[_TT_Sk]** A *transport processing function* that accepts the *characteristic information* of the layer network at its input, removes the information related to *trail* monitoring and presents the remaining *adapted information* at its output to the *adaptation sink* function. A trail termination sink function can operate without an output to a client layer network.
trail termination source	**[_TT_So]** A *transport processing function* that accepts *adapted information* from a client layer network at its input, adds information to allow the *trail* to be monitored and presents the *characteristic information* of the layer network at its output. A trail termination source function can operate without an input from a client layer network.
transmission	The physical process of propagating information signals through a physical medium.

Term	Description/definition
transmission layer network	A *layer network* that may be media-dependant and that is concerned with the transfer of information between *section layer access points* in support of one or more *path layer networks*. It is further divided into a *section layer network* and a *physical media layer network*. It is also known as transmission media layer network.
transport	The functional process of transferring information between different locations.
transport assembly	An arbitrary combination of contiguous *layer networks* and *adaptation functions*.
transport entity	An *architectural component* that transfers information between its inputs and outputs within a *layer network*.
transport network	The set of functional resources in a network that conveys user information between locations. It is built by using successive *transport network layers*, one upon another. Each *transport network layer* provides *transport* for the (client) layer above and uses *transport* provided by the (server) layer below.
transport network layer	Defined, at its highest level, by the *trails* that it supports or is capable of supporting and is characterized by its *characteristic information*.
transport processing function	An architectural component defined by the information processing that is performed between its inputs and outputs. Either the input or output must be inside a layer network; the corresponding output or input may be in the Management Network (e.g. output of a monitor function).
tributary unit	(TU) The information structure that is used to provide the *adaptation* between the lower order path layer and the higher order path layer. It consists of an information payload (the lower order VC) and a TU pointer that indicates the offset of the payload frame start relative to the higher order VC frame start.
uni-directional access point	A *reference point* where the output of a *trail* or *flow termination sink* is bound to the input of an *adaptation sink* function or the output of an *adaptation source* function is bound to an input of a *trail* or *flow termination source*.

Term	Description/definition
	The *access point* is characterized by the *adapted information* that passes across it.
uni-directional connection	A *transport entity* that transfers information transparently from input to output.
uni-directional connection point	A *reference point* that represents the binding of either the output of a *connection source* function to the input of an *adaptation source* function, or the output of an *adaptation sink* function to the input of a *connection sink* function.
uni-directional port	It represents either the input to a *trail termination source*, a uni-directional link connection, a *flow termination source*, or a *flow domain*, or the output of a *trail termination sink*, a uni-directional link connection, a *flow termination sink*, or a *flow domain*.
uni-directional termination connection point	A reference point that represents the binding of an output of a *trail termination source* to the input of a *uni-directional connection*, or the output of a *uni-directional connection* to the input of a *trail termination sink*.
uni-directional trail	A *transport entity* responsible for the transfer of information from the input of a *trail termination source* to the output of a *trail termination sink*. The integrity of the information transfer is monitored. It is formed by combining *trail termination functions* and a network connection.
user network interface	(UNI) The interface at a network node that is used to interconnect with a user.
virtual container	(VC) The information structure used to support path layer connections in the SDH. It consists of information payload and path overhead information fields organized in a block frame structure that repeats every 125 or 500 μsec. Alignment information to identify VC frame start is provided by the server network layer.

References

(AMND. = AMENDMENT, CORR. = CORRIGENDUM, ERRT. = ERRATUM)

ANSI T1.105 (2001), *Synchronous Optical Network (SONET) – Basic description including multiplex structure, rates, and formats.*

S. Brown, 'A Functional Description of SDH Transmission Equipment', *BT Technology Journal*, Volume 14, No. 2, 1996.

ETSI EN 300 417-1-1 v1.2.1 (10/2001), *Generic requirements of transport functionality of equipment; Part 1–1: Generic processes and performance.*

ETSI EN 300 417-2-1 v1.2.1 (10/2001), *Generic requirements of transport functionality of equipment; Part 2–1: Synchronous Digital Hierarchy (SDH) and Plesiochronous Digital Hierarchy (PDH) physical section layer functions.*

ETSI EN 300 417-3-1 v1.2.1 (10/2001), *Generic requirements of transport functionality of equipment; Part 3–1: Synchronous Transport Module-N (STM-N) regenerator and multiplex section layer functions.*

ETSI EN 300 417-4-1 v1.2.1 (10/2001), *Generic requirements of transport functionality of equipment; Part 4–1: SDH path layer functions.*

ETSI EN 300 417-5-1 v1.2.1 (10/2001), *Generic requirements of transport functionality of equipment; Part 5–1: Plesiochronous Digital Hierarchy (PDH) path layer functions.*

ETSI EN 300 417-6-1 v1.1.3 (05/1999), *Generic requirements of transport functionality of equipment; Part 6–1: Synchronization layer functions.*

ETSI EN 300 417-7-1 v1.1.1 (10/2000), *Generic requirements of transport functionality of equipment; Part 7–1: Equipment management and auxiliary layer functions.*

ETSI EN 300 417-9-1 v1.1.1 (09/2001), *Generic requirements of transport functionality of equipment; Part 9–1: SDH concatenated path layer functions; Requirements.*

SDH/SONET Explained in Functional Models Huub van Helvoort

ETSI EN 300 417-10-1 v1.1.1 (11/2003), *Generic requirements of transport functionality of equipment; Part 10–1: Synchronous Digital Hierarchy (SDH) radio specific functionalities.*

IEEE Std. 802 (2001), IEEE Standard for Local and Metropolitan Area Networks: Overview and Architecture.

IEEE Std 802.1D (2004), IEEE Standard for Local and Metropolitan Area Networks: Media Access Control (MAC) Bridges.

IEEE Std 802.1Q (2003), IEEE Standard for Local and Metropolitan Area Networks: Virtual Bridged Local Area Networks.

IEEE Std. 802.3 (2002), IEEE Standard for Information Technology – Telecommunications and information exchange between systems – IEEE standard for local and metropolitan area networks – Specific requirements – Part 3: *Carrier Sense Multiple Access with Collision Detection (CSMA/CD) Access Method and Physical Layer Specifications.*

IEEE Std 802-3ae (2002), Information technology – Telecommunications and information exchange between systems – Local and metropolitan area networks – Specific requirements – Part 3: *Carrier Sense Multiple Access with Collision Detection (CSMA/CD) Access Method and Physical Layer Specifications.* Amendment: Media Access Control (MAC) Parameters, Physical Layers, and Management Parameters for 10 Gbit/s Operation.

ITU-T Recommendation G.702 (11/1988), *Digital Hierarchy bit rates.*

ITU-T Recommendation G.703 (1998), *Physical/electrical characteristics of hierarchical digital interface.*

ITU-T Recommendation G.704 (10/1998), *Synchronous frame structures used at 1544, 6312, 2048, 8448 and 44 736 kbit/s hierarchical levels.*

ITU-T Recommendation G.705 (10/2000), *Characteristics of plesiochronous digital hierarchy (PDH) equipment functional blocks.*

ITU-T Recommendation G.706 (04/1991), *Frame alignment and cyclic redundancy check (CRC) procedures relating to basic frame structures defined in Recommendation G.704.*

ITU-T Recommendation G.707 (12/2003), *Network node interface for the Synchronous Digital Hierarchy SDH*, [Amnd. 1 08/2004, Corr. 1 06/2004].

ITU-T Recommendation G.709 (03/2003), *Network node interface for the Optical Transport Network (OTN)*, [Amnd. 1 12/2003].

ITU-T Recommendation G.774 (02/2001), *Synchronous digital hierarchy (SDH) – Management information model for the network element view.*

ITU-T Recommendation G.781 (07/1999), *Synchronization layer functions*, [Corr. 1 06/2004].

ITU-T Recommendation G.783 (02/2004), *Characteristics of Synchronous Digital Hierarchy (SDH) equipment functional blocks*, [Corr. 1 06/2004].

ITU-T Recommendation G.784 (07/1999), *Synchronous Digital Hierarchy (SDH) management.*

ITU-T Recommendation G.798 (06/2004), *Characteristics of optical transport network equipment functional blocks.*

ITU-T Recommendation G.7041/Y.1303 (12/2003), *Generic Framing Procedure (GFP)*, [Amnd. 1 10/2004, Amnd. 2 06/2004, Amnd. 3 01/2005, Corr. 1 01/2005].
ITU-T Recommendation G.7042/Y.1305 (02/2004), *Link Capacity Adjustment Scheme (LCAS) for virtual concatenated signals*, [Corr. 1 08/2004].
ITU-T Recommendation G.7710 (11/2001), *Common equipment management function requirements*.
ITU-T Recommendation G.803 (03/2000), *Architecture of transport networks based on the synchronous digital hierarchy (SDH)*.
ITU-T Recommendation G.805 (03/2000), *Generic functional architecture of transport networks*.
ITU-T Recommendation G.806 (02/2004), *Characteristics of transport equipment – description methodology and generic functionality*, [Amnd. 1 06/2004, Corr. 1 08/2004, Corr. 2 01/2005].
ITU-T Recommendation G.808.1 (12/2003), *Generic protection switching – Linear trail and subnetwork protection*.
ITU-T Recommendation G.809 (03/2003), *Functional architecture of connectionless layer networks*.
ITU-T Recommendation G.810 (08/1996), *Definitions and terminology for synchronization networks*, [Corr. 1 11/2001].
ITU-T Recommendation G.811 (09/1997), *Timing characteristics of primary reference clocks*.
ITU-T Recommendation G.812 (06/2004), *Timing requirements of slave clocks suitable for use as node clocks in synchronization networks*.
ITU-T Recommendation G.813 (03/2003), *Timing characteristics of SDH equipment slave clocks (SEC)*.
ITU-T Recommendation G.822 (11/1988), *Controlled slip rate objectives on an international digital connection*.
ITU-T Recommendation G.841 (10/1998), *Types and characteristics of SDH network protection architectures*, [Corr. 1 08/2002].
ITU-T Recommendation G.842 (04/1997), *Interworking of SDH network protection architectures*.
ITU-T Recommendation G.872 (11/2001), *Architecture of optical transport networks*, [Amnd. 1 12/2003, Corr. 1 01/2005].
ITU-T Recommendation G.873.1 (03/2003), *Optical Transport Network (OTN): Linear protection*, [Errt. 1 10/2003].
ITU-T Recommendation G.874 (11/2001), *Management aspects of the optical transport network element*.
ITU-T Recommendation G.8010/Y.1306 (02/2004), *Architecture of Ethernet layer networks*.
ITU-T Recommendation G.8011/Y.1307 (08/2004), *Ethernet over Transport–Ethernet services framework*.
ITU-T Recommendation G.8011.1/Y.1307.1 (08/2004), *Ethernet private line service*.
ITU-T Recommendation G.8012/Y.1308 (08/2004), *Ethernet UNI and Ethernet over transport NNI*.

ITU-T Recommendation G.8021/Y.1341 (08/2004), *Characteristics of Ethernet transport network equipment functional blocks.*

ITU-T Recommendation G.8080/Y.1304 (2001), *Architecture for the automatic switched optical network (ASON).*

ITU-T Recommendation G.8110/Y.1370 (01/2005), *MPLS layer network architecture.*

ITU-T Recommendation G.mplseq (03/2005), *Characteristics of Multi Protocol Label Switched (MPLS) equipment functional blocks*, [draft 0.2].

ITU-T Recommendation I.326 (1995), *Functional architecture of transport networks based on ATM.*

ITU-T Recommendation I.732 (10/2000), *Functional characteristics of ATM equipment.*

ITU-T Recommendation M.3100 (07/1995), *Generic network information model*, [Amnd. 1 03/1999, Amnd. 2 02/2000, Amnd. 3 01/2001, Amnd. 4 08/2001, Amnd 5 08/2001, Amnd 6 03/2003, Amnd 7 12/2003, Amnd 8 08/2004, Corr. 1 06/1998, Corr. 2 01/2001, Corr. 3 08/2001].

ITU-T Recommendation O.150 (05/2002), *General requirements for instrumentation for performance measurements on digital transmission equipment*, [Corr. 1 05/2002].

ITU-T Recommendation X.200 (07/1994), *Information technology – Open Systems Interconnection – Basic Reference Model: The basic model.*

ITU-T Recommendation Y.1710 (11/2002), *Requirements for OAM functionality for MPLS networks.*

ITU-T Recommendation Y.1730 (01/2004), *Requirements for OAM functions in Ethernet-based networks and Ethernet services.*

ITU-T Recommendation Z.100 (08/2002), *Specification and Description Language* SDL, [Amnd. 10/2003, Corr. 08/2004].

ITU-T Recommendation Z.200 (11/1999), *CHILL – The ITU-T Programming Language.*

Brian W. Kernighan and Dennis M. Ritchie (1978), *The C programming Language*, Prentice-Hall.

FURTHER READING

Huub van Helvoort (2005), *Next Generation SDH/SONET: Evolution or Revolution*, John Wiley & Sons, Ltd.

Mike Sexton and Andy Reid (1997), *Broadband Networking: ATM, SDH, and SONET*, Artech House.

Eve Varma, George Newsome, *et al.* (1999), *Achieving Global Information Networking*, Artech House.

Index